A Modern Hand Book of Environmental Economics

A Modern Hand Book of Environmental Economics

S.S. Rana

CYBER TECH PUBLICATIONS
4264/3, Ansari Road, Daryaganj, New Delhi-110002 (India)
Ph.: 011-23244078, 011-43559448 Fax: 011-23280028
E-Mail: cyberpublicationsdelhi@yahoo.com
Website: www.cybertechpublications.com

A Modern Hand Book of Environmental Economics

S.S. Rana

First Edition 2010

Asian Offset Printers

Published by :

G.S. Rawat for **Cyber Tech Publications**
4264/3, Ansari Road, Daryaganj, New Delhi-110002 (India)
Ph.: 011-23244078, 011-43559448 Fax: 011-23280028
E-Mail: cyberpublicationsdelhi@yahoo.com
Website: www.cybertechpublications.com

Preface

In this book "A Modern Hand Book of Envrionmental Economics, almost half the world's six billion people live on two dollars a day or less. Between one third and one half suffer undernutrition due to insufficient intake of calories. More than one child in five lives an especially disturbing fact to this father of five. Why does such unnecessary injustice continue to disfigure a rich, technologically advanced world and what can be done to care for the poor and thereby to care for and honor God, as the Gospels instruct us?

We are called to feed the hungry, nurse the sick, and clothe the naked. The impulse to assist is obviously not unique to Christendom. In reflecting on the economics of poverty, my focus is therefore not on how best to run humanitarian operations but, rather, on the big picture mechanisms that necessitate the grim but honorable and too-necessary work of humanitarian relief agencies.

This essay was composed during a fascinating country in which extraordinarily high species endemism rates fuel intense interest by conservationists and where the unique blend of Polynesian and Bantu cultures has long drawn the attention of cultural anthropologists. Such obvious unique qualities aside, Madagascar is nonetheless the quintessential poor economy.

— *Author*

Preface

CONTENTS

1

BASIC CONCEPTS OF THE ENVIRONMENTAL ECONOMICS

INTRODUCTION

Environmental economics is a subfield of economics concerned with environmental issues. Environmental Economics undertakes theoretical or empirical studies of the economic effects of national or local environmental policies around the world. Particular issues include the costs and benefits of alternative environmental policies to deal with air pollution, water quality, toxic substances, solid waste, and global warming. The environmental economics is the concept of market failure.

Market failure means that markets fail to allocate resources efficiently. "A market failure occurs when the market does not allocate scarce resources to generate the greatest social welfare. A wedge exists between what a private person does given market prices and what society might want him or her to do to protect the environment. Such a wedge implies wastefulness or economic inefficiency; resources can be reallocated to make at least one person better off without making anyone else worse off." Common forms of market failure include externalities, non excludability and non rivalry.

PROFESSIONAL BODIES

The main academic and professional organizations for the discipline of Environmental Economics are the Association of Environmental and Resource Economists (AERE) and the European Association for Environmental and Resource Economics (EAERE). The main academic and professional organization for the discipline of Ecological Economics is the International Society for Ecological Economics (ISEE

COMMON PROPERTY AND NON-EXCLUSION

When it is too costly to exclude people from accessing a rivalrous environmental resource, market allocation is likely to be inefficient. The challenges related with common property and non-exclusion have long been recognized. Hardin's (1968) concept of the tragedy of the commons popularized the challenges involved in non-exclusion and common property. "commons" refers to the environmental asset itself, "common property resource" or "common pool resource" refers to a property right regime that allows for some collective body to devise schemes to exclude others, thereby allowing the capture of future benefit streams; and "open-access" implies no ownership in the sense that property everyone owns nobody owns. The basic problem is that if people ignore the scarcity value of the commons, they can end up expending too much effort, over harvesting a resource (e.g., a fishery). Hardin theorizes that in the absence of restrictions, users of an open-access resource will use it more than if they had to pay for it and had exclusive rights, leading to environmental degradation. See, however, Ostrom's (1990) work on how people using real common property resources have worked to establish self-governing rules to reduce the risk of the tragedy of the commons.

PUBLIC GOODS AND NON-RIVALRY

Public goods are another type of market failure, in which the market price does not capture the social benefits of its provision. For example, protection from the risks of climate change is a public good since its provision is both non-rival and non-excludable. Non-rival means climate protection provided to one country does not reduce the level of protection to another country; non-excludable means it is too costly to exclude any one from receiving climate protection. A country's incentive to invest in carbon abatement is reduced because it can "free ride" off the efforts of other countries. Over a century ago, Swedish economist Knut Wicksell (1896) first discussed how public goods can be under-provided by the market because people might conceal their preferences for the good, but still enjoy the benefits without paying for them.

EXTERNALITY

The basic idea is that an externality exists when a person makes a choice that affects other people that are not accounted for in the market price. For instance, a firm emitting pollution will typically not take into account the costs that its pollution imposes on others. As a result, pollution in excess of the 'socially efficient' level may occur. A classic definition is provided by Kenneth Arrow (1969), who defines an externality as "a situation in which a private economy lacks sufficient incentives to create a potential market in some good, and the nonexistence of this market results in the loss of efficiency." In economic terminology, externalities are examples of market failures, in which the unfettered market does not lead to an efficient outcome.

ENVIRONMENTAL IMPACT

"Environmental Impact Assessment can be defined as: The process of identifying, predicting, evaluating and mitigating the biophysical, social, and other relevant effects of development proposals prior to major decisions being taken and commitments made." The purpose of the assessment is to ensure that decision-makers consider environmental impacts before deciding whether to proceed with new projects.

AN OVERVIEW

The EPA employs 17,000 people in headquarters program offices, 10 regional offices, and 27 laboratories across the country. More than half of its staff are engineers, scientists, and environmental protection specialists; other groups include legal, public affairs, financial, and computer specialists.

The agency conducts environmental assessment, research, and education. It has the primary responsibility for setting and enforcing national standards under a variety of environmental laws, in consultation with state, tribal, and local governments. It delegates some permitting, monitoring, and enforcement responsibility to U.S. states and Native American tribes. EPA enforcement powers include fines, sanctions, and other measures. The agency also works with industries and all levels of government in a wide variety of voluntary pollution prevention programs and energy conservation efforts.

The technology for performing such analysis is properly labelled environmental science. The principal phenomena or pathways of impact are: soil contamination impacts, air pollution impacts, noise health effects, ecology impacts including endangered species assessment, geological hazards assessment and water pollution impacts. Pathway analysis and The Natural Step definitions subsequently became the basis of the global ISO 14000 series of environmental management standards and the more recent ISO 19011 auditing standard; however, these ISO standards are not in common use in the U.S. and most other countries.

The US Environmental Protection Agency pioneered the use of pathway analysis to determine the likely human health impact of environmental factors. After an EIA analysis, the Precautionary Principle and Polluter Pays may be applied to prevent, limit, or require strict liability or insurance coverage to a project, based on its likely harms. Environmental impact analysis is sometimes controversial and contested. Related analysis of social impacts is achieved by Social impact assessment. Analysis of business impacts is achieved by context analysis. Design impacts are assessed in relation to context theory. EIA predicts what a specific action can do to the environment.

China

The Environmental Impact Assessment Law (EIA Law) requires an environmental impact assessment to be completed prior to project construction. However, if a developer completely ignores this requirement and builds a project without submitting an environmental impact statement, the only penalty is that the environmental protection bureau (EPB) may require the developer to do a make-up environmental assessment.

If the developer does not complete this make-up assessment within the designated time, only then is the EPB authorized to fine the developer. Even so, the possible fine is capped at a maximum of about US$25,000, a fraction of the overall cost of most major projects. The lack of more stringent enforcement mechanisms has resulted in a significant percentage of projects not completing legally required environmental impact assessments prior to construction.

China's State Environmental Protection Administration (SEPA) used the legislation to halt 30 projects in 2004, including three hydro-power plants under the Three Gorges Project Company. Although one month later (Note as a point of reference, that the typical EIA for a major project in the USA takes one to two years.), most of the 30 halted projects resumed their construction, reportedly having passed the environmental assessment, the fact that these key projects' construction was ever suspended was notable.

A joint investigation by SEPA and the Ministry of Land and Resources in 2004 showed that 30 to 40 per cent of the mining construction projects went through the procedure of environment impact assessment as required, while in some areas only 6 to 7 per cent did so.

This partly explains why China has witnessed so many mining accidents in recent years. SEPA alone cannot guarantee the full enforcement of environmental laws and regulations, observed Professor Wang Canfa, director of the centre to help environmental victims at China University of Political Science and Law. In fact, according to Wang, the rate of China's environmental laws and regulations that are actually enforced is estimated to be barely 10 per cent.

EU

The EIA Directive on Environmental Impact Assessment of the effects of projects on the environment was first introduced in 1985 and was amended in 1997. The directive was amended again in 2003 following the 1998 signature by the EU of the Aarhus Convention on public participation in environmental matters.

The issue was enlarged to the assessment of plans and programmes by the so called SEA-Directive in 2001 which is now in force and establishes a mix of mandatory and discretionary procedures for assessing environmental impacts. Under the EU directive, an EIA must provide certain information to comply. There are seven key areas that are required:

Description of the Project

- Description of actual project and site description
- Break the project down into its key components, ie construction, operations, decommissioning
- For each component list all of the sources of environmental disturbance
- For each component all the inputs and outputs must be listed, eg, air pollution, noise, hydrology

Alternatives that have been Considered

- Examine alternatives that have been considered
- Example: in a biomass power station, will the fuel be sourced locally or nationally?

Description of the Environment

- List of all aspects of the environment that may be effected by the development
- Example: populations, fauna, flora, air, soil, water, humans, landscape, cultural heritage
- This section is best carried out with the help of local experts, eg the RSPB in the UK

Description of the Significant Effects on the Environment

- The word significant is crucial here as the definition can vary
- 'Significant' needs to be defined
- The most frequent method used here is use of the Leopold matrix
- The matrix is a tool used in the systematic examination of potential interactions
- Example: in a windfarm development a significant impact may be collisions with birds

Mitigation

- This is where EIA is most useful
- Once section 4 has been completed it will be obvious where the impacts will be greatest
- Using this information ways to avoid negative impacts should be developed
- Best working with the developer with this section as they know the project best
- Using the windfarm example again construction could be out of bird nesting seasons

Non-technical Summary

- The EIA will be in the public domain and be used in the decision making process
- It is important that the information is available to the public
- This section is a summary that does not include jargon or complicated diagrams
- It should be understood by the informed lay-person

Lack of Know-how/technical Difficulties

- This section is to advise any areas of weakness in knowledge
- It can be used to focus areas of future research
- Some developers see the EIA as a starting block for good environmental management

New Zealand

In New Zealand, EIA is usually referred to as Assessment of Environmental Effects (AEE). The first use of EIA's dates back to a Cabinet minute passed in 1974 called Environmental Protection and Enhancement Procedures. This had no legal force and only related to the activities of government departments. When the Resource Management Act was passed in 1991, an EIA was required as part of a resource consent application. Section 88 of the Act spells this out.

United States

Under United States environmental law an EIA is referred to as the Environmental Impact Statement (EIS), and originated in the National Environmental Policy Act (NEPA), enacted in the United States in 1969. Certain actions of federal agencies must be preceded by an EIS. Contrary to a widespread misconception, NEPA does not prohibit the federal government or its licensees/permittees from harming the environment, nor does it specify any penalty if the EIS turns out to be inaccurate, intentionally or otherwise. NEPA requires that plausible statements as to the prospective impacts be disclosed in advance. The purpose of NEPA process is to ensure that the decision maker is fully informed of the environmental aspects and consequences prior to making the final decision.

Usually, an agency will release a Draft Environmental Impact Statement (DEIS) for comment. Interested parties and the general public have the opportunity to comment on the draft, after which the agency will approve the Final Environmental Impact Statement (FEIS). Occasionally, the agency will later release a Supplemental Environmental Impact Statement (SEIS), especially if environmental conditions suddenly change after the issuance of the FEIS. The adequacy of an EIS can be challenged in court. Major proposed projects have been blocked because of an agency's failure to prepare an acceptable EIS. One prominent example was the Westway landfill and highway development in and along the Hudson River in New York City.

Another prominent case involved the Sierra Club suing the Nevada Department of Transportation over its denial of Sierra Club's request to issue a supplemental EIS addressing air emissions of particulate matter and hazardous air pollutants in the case of widening US Highway 95 through Las Vegas. The case reached the 9th Circuit Court of the United States, which led to construction on the highway being halted until the court's final decision. The case was settled prior to the court's final decision.

Several US state governments that have adopted "little NEPA's," i.e., state laws imposing EIS requirements for particular state actions and some of those state laws refer to the required environmental impact studies as Environmental Impact Reports or Environmental Impact Assessments. For example, the California Environmental Quality Act (CEQA) requires an Environmental Impact Report (EIR).

These various state requirements are yielding voluminous data not just upon impacts of individual projects, but also to elucidate scientific areas that had not been sufficiently researched. For example, in a seemingly routine Environmental Impact Report for the city of Monterey, California, information came to light that led to the official federal endangered species listing of Hickman's potentilla, a rare coastal wildflower.

ENVIRONMENTAL IMPACT ASSESSMENT

The current Regulations are The Environmental Impact Assessment (Forestry) (England and Wales) Regulations 1999 [SI 1999/2228] and the Environmental Impact Assessment (Forestry) (Scotland) Regulations 1999 [SI 1999/43]. These Regulations have been further amended by

The Environmental Impact Assessment (Forestry) (England and Wales) (Amendment) Regulations 2006 and The Environmental Impact Assessment (Scotland) Amendment Regulations 2006. These Regulations came into force on 6th September 1999 and require anyone who wishes to carry out a relevant project (i.e. deforestation, afforestation, forestry roads or quarries that might have a significant effect on the environment) to obtain consent for the work from the Forestry Commission. The applicant or proposer must submit an Environmental Statement in support of the proposals to apply for consent.

You will find below some of the terms used in the EIA Regulations:

a) Appropriate Authority-The Secretary of State for Environment, Food and Rural Affairs in England, the National Assembly for Wales, and the Scottish Executive in Scotland.

b) Countryside bodies

   England-Natural England, Environment Agency. Wales-Countryside Council for Wales, Environment

Agency. Scotland-Scottish Natural Heritage, Scottish Environmental Protection Agency. And any other body designated by statutory provision as having specific environmental responsibilities.

c) Determination-Taken from Regulation 15 "Determination of applications" and is the process by which we make our decision about the application for consent.

d) Forestry projects-the project work that the Forestry Commission must assess under these Regulations are deforestation (conversion to another land use), afforestation, forestry roads and forestry quarries.

e) Opinion-Our consideration of the proposals from which we will decide whether or not the project is a relevant one. If it is, the applicant must apply for consent and provide an ES.

f) Relevant project-A forestry project (afforestation, deforestation, forest roads works and forest quarry works) that is likely, by virtue of factors such as its nature, size and location, to have a significant effect on the environment and as such requires the FC's consent.

g) Scoping-A gathering of all consultees and other interested parties to discuss and agree the significant issues of concern that require to be addressed by an applicant when preparing an Environmental Statement.

h) Screening-the process by which the Forestry Commission decides whether a project "is likely to have significant effects on the environment by virtue, inter alia, of its nature, size or location". This process is a distinct one from scoping.

i) Publicity-The Forestry Commission maintains a web-based EIA Register that gives details of all the decisions we make under these Regulations. Applicants are also required to advertise details of any application for consent.

j) Thresholds-Area limits set by the Regulations below which it is not expected that the project will have a significant effect on the environment.

The Forestry Commission will assess whether:

- o the proposed project is one of the following categories- afforestation, deforestation, forest roads or quarries;
- o the area is above the relevant threshold (includes extensions to similar areas of work);
- o the project is likely to have a significant effect on the environment.

Our consent will be required to proceed with the work if your proposals meet each of these requirements. You will be asked to provide an Environmental Statement to allow us to decide whether to give consent to the project. Under the 1999 Regulations, proposals are considered to be relevant projects (i.e. to require an EIA) if they fall within the categories listed below and the work proposed is likely to have a significant effect on the environment.

a) Initial afforestation: creating new woods and forests by planting trees (on an area that has not had trees for many years). This category includes using direct seeding or natural regeneration, planting Christmas trees and short rotation coppice;

b) Deforestation: conversion of woodland to another type of land use (e.g.heathland);

c) Forest roads: constructing forestry roads, including those within a forest and those leading to a forest;

d) Forest quarries: quarrying to obtain material (rock, sand and gravel) for the formation, alteration or maintenance of forest roads.

We may serve you with an Enforcement Notice if you carry out work on a project that would have required our consent. This notice will require you to comply with the Regulations. Situations may arise where one of the above forestry projects forms part of a wider development that requires Planning Permission.

In these circumstances, any necessary EIA will usually not be dealt with under the Forestry EIA Regulations but under the parallel Town and Country Planning EIA Regulations.

Where a project has been completed since 6 September 1999, work proposals of the same type (including that on land in different ownerships) that extend it beyond the thresholds may need our consent. This also applies where the accumulated area of all adjacent existing projects plus the new proposals, exceeds the thresholds.

Making the decision Definition of relevant projects Proposals that extend a project area previously below the area thresholds

If the area of the work you are proposing falls below these thresholds then it is not likely to be a relevant project and our consent under the EIA Regulations will not be required. You may still of course require a felling licence or to get approval under the Woodland Grant Scheme for the work.

If your project is above these area thresholds, our consent for carrying out the work may be required.

You must provide the following so that we can give our opinion on whether the work you are proposing is a relevant project and requires our consent:

a) a map identifying the area and showing the extent of the project. (This should be a clear Ordnance Survey map at a scale of 1:10 000 or 1:2 500);

b) a brief description of the proposed work and its possible significant effects on the environment;

c) any other information that might be relevant, e.g. species maps, plans, photographs etc.

In the majority of instances this information will be provided as part of a Woodland Grant Scheme application (initial afforestation) or an application for a felling licence (deforestation) or Forest Plan.

If, at any time, we become aware of proposals that require our consent, we will give our opinion to the person whom we believe should have asked us. We will give our opinion, in writing, within 28 days of receiving the necessary information. However, if we need more information from you to help us decide, then we will let you have a decision within 28 days of receiving this information. If our decision will take longer we will seek to agree an appropriate period with you. If we tell you that your proposal is a relevant project, or we do not give you our opinion within the agreed timescale, you can appeal in writing to: The Scottish Ministers (in relation to Scotland); The Secretary of State for Environment, Food and Rural Affairs; The National Assembly for Wales in Wales for a "direction" as to whether the work is a relevant project. Once we have told you our decision we will publish this on an "EIA Public Register" held at Forestry Commission offices. This will show information about our opinion whether a project is a relevant project, our decision about the granting of consent and any directions that may have been made by the appropriate authority or the Scottish Ministers.

Getting our Consent

We must have the following information so that we can make a decision:-

a) a map showing the area where the project is proposed. (This should be a clear Ordnance Survey map at a scale of 1:10 000 or 1:2 500);

b) a brief description of the proposed work and the possible significant effects on the environment;

c) any other information that might be relevant e.g. species maps, plans, photographs etc.

(You may already have supplied sufficient information for the above if you have asked for our Opinion under the EIA regulations or submitted a Woodland Grant Scheme or Forest Plan application)

d) an Environmental Statement for the work;

e) a copy of the publicity notice that you must place in newspapers.

Before you start to prepare your ES, we would encourage you to meet us and discuss the significant environmental issues that will need to be covered, or ask for information to help you. At an early stage, it is strongly recommended that you contact the appropriate countryside bodies* and request any information they may have that is relevant to the project. [Please note that there may be a small charge for providing this information].

* In England-Natural England, English Heritage, Environment Agency; In Wales-Countryside Council for Wales, Environment Agency;

In Scotland-Scottish Natural Heritage, Scottish Environmental Protection Agency

In addition we would advise that you hold a scoping meeting with countryside bodies and other relevant organisations to help determine the issues that must be covered in the Environmental Statement. There is no set format for an ES but it must contain the standard information contained in Appendix 2. Your Statement must include everything from the list that relates to your application, placing the emphasis on the significant environmental effects. You can also get more advice about preparing an Environmental Statement in our booklet-"Preparing an Environmental Statement for a Forestry Project", only available as a PDF download from www.forestry.gov.uk.

It may help to facilitate your application if you let us have a draft copy of your ES before preparing the final copy. This will help us to decide if you have covered and addressed all the relevant issues. If we need more information to help us decide about your application, you must supply this. You can get a copy of this booklet on the Forestry Commission Grants and Licences section of the website. If you have applied for our consent and prepared an ES, you must publish a notice giving the details outlined below. We will agree at least two newspapers in which the notice must appear.

The notice must:

a) state that an application for our consent under the Regulations has been made;

b) indicate where (and at what times) copies of the application for consent can be inspected (the local Forestry Commission office or other location(s) we agree);

c) state the address where people can get copies of the application;

d) give details of any charge that may be made for copies-to cover printing, copying and distribution costs only;

e) state that comments about the application must be made to the relevant Forestry Commission Conservator within 28 days of the date of the notice.

f) You must also include the following statement: "The Forestry Commissioners may decide either to grant consent subject to the standard conditions required by Regulation 18 of the EIA Regulations, or subject to such further conditions as they see fit or refuse consent."

We will also give details of your application to the appropriate countryside body as well as the local authority with an interest in the application. They are required to give us their comments within 28 days.

We will give you our decision (and the reasons for it) in writing, once the 28-day period for comments has expired and we have considered the environmental information, any representations and any other material consideration. A notice giving details of our decision will be published in the same newspapers in which the initial notice about the application appeared. We will also inform anyone from whom we have received representations in relation to the application.

When granting consent, we will specify the date by which you must start and finish the work. The start date will be no later than five years from the date of the consent and the finish date will be no later than ten years from the date of the consent.

We may also apply other conditions to the consent that we deem necessary to ensure that the environment is protected from the impact of the project. These may be in addition to the normal Grant Scheme or Felling Licence conditions.

Publicity about the application Consultation about the application Giving our decision about your application Standard consent conditions.

If you have applied for consent, you can appeal about the decision we make to:

*The Scottish Ministers (in relation to Scotland);

*The Secretary of State for Environment, Food and Rural Affairs in England;

*The National Assembly for Wales in Wales;

You must do this within 28 calendar days of receiving notice of our decision or within any other period that the Government authority* may agree.

You may only appeal if the Forestry Commissioners have:

a) refused the application;

or

b) have granted consent with conditions (additional to the standard conditions); or

c) we have asked you to start and finish the work within a period that is less than the maximum period permitted by the Regulations. In addition, any person who feels aggrieved by the granting of consent can make an application to the Court if they feel that their interests are prejudiced. They must do this within 6 weeks of consent being granted.

If we discover that you have carried out work that would have required our consent, or have breached the conditions of consent, we may serve you with an Enforcement Notice. This will require you to do one or more of the following:

a) stop the work;

b) apply for consent;

c) restore the land to its condition before the work was started;

d) carry out work to secure compliance with the conditions of the consent;

e) remove or alleviate any damage to the environment that has been caused by the work.

You can appeal against the serving of the Enforcement Notice to:

*The Scottish Ministers (in relation to Scotland);

*The Secretary of State for Environment, Food and Rural Affairs in England;

*The National Assembly for Wales in Wales;

You must do this within 28 calendar days of receiving notice of our decision or within any other period that the Government authority* may agree.

If you require more information or help, including how to make appeals, contact your local Forestry Commission office or: Grants and Licences, Forestry Commission Headquarters, 231 Corstorphine Road, Edinburgh EH12 7AT

Appealing against our decision Carrying out work without our consent or breaching the conditions of consent Appealing against the enforcement notice Getting more information.

Definition of "sensitive areas":

For the purposes of the the definitions of "sensitive areas" are as follows:

a) National Nature Reserve or Site of Special Scientific Interest;

b) National Parks;

c) The Broads;

d) A World Heritage Site;

e) Scheduled Ancient Monuments;

f) Area of Outstanding Natural Beauty;

g) Natural Heritage or National Scenic Areas;

h) A site designated or identified as a candidate Special Area of Conservation;

i) The New Forest Heritage Area;

j) A site classified or proposed as a Special Protection Area.

Information for inclusion in environmental statements Part 1.

1) Description of the project, including in particular:

   a) a description of the physical characteristics of the whole project and the land-use requirements during the construction and operational phases;

   b) a description of the main characteristics of the production processes, for instance, nature and quantity of the materials used;

c) an estimate, by type and quantity, of expected residues and emissions (water, air and soil pollution, noise, vibration, light heat, radiation, etc) resulting from the operation of the proposed project.

2) An outline of the main alternatives studied by the developer and an indication of the main reasons for the developer's choice, taking into account environmental effects.

3) A description of the aspects of the environment likely to be significantly affected by the proposed project, including, in particular, human beings, fauna, flora, soil, water, air, climatic factors, material assets including the architectural and archaeological heritage, landscape and the inter-relationship between the above factors.

4) A description of the likely significant effects of the proposed project on the environment, which should cover the direct effects and any indirect, secondary, cumulative, short, medium and long-term, permanent and temporary, positive and negative effects of the project, resulting from:

a) the existence of the project;

b) the use of natural resources;

c) the emission of pollutants, the creation of nuisances, and the elimination of waste, and the description by the developer of the forecasting methods used to assess the effects on the environment.

5) A description of the measures proposed to prevent, reduce and, where possible offset, any significant adverse effects on the environment.

6) A non-technical summary of the information provided under paragraphs 1 to 5 above.

7) An indication of any difficulties (technical deficiencies or lack of knowledge) encountered by the developer in compiling the required information.

GUIDE TO PROCEDURES

Environmental impact assessment (EIA) is an important procedure for ensuring that the likely effects of new development on the environment are fully understood and taken into account before the development is allowed to go ahead. This booklet, which is intended primarily for developers and their advisers, explains how European Community (EC) requirements for the environmental impact assessment of major projects have been incorporated into consent procedures in the UK.

It revises the booklet 'Environmental Assessment: A Guide to the Procedures', first published in 1989, to take account of the requirements of Directive 97/11/EC, which was adopted on 3 March 1997 and came into effect on 14 March 1999. Directive 97/11/EC amends the original Directive 85/337/EEC on 'The assessment of the effects of certain public and private projects on the environment', which came into effect in July 1988. The text of the Directive as amended (with amendments shown in italics) is reproduced in Appendix 1 to this booklet. Throughout this booklet, references to `the Directive' mean Directive 85/337/EEC as amended by Directive 97/11/EC.

The effect of the Directive is to require environmental impact assessment to be carried out, before development consent is granted, for certain types of major project which are judged likely to have significant environmental effects. Parts 1 and 2 of this booklet explain the procedures which apply to projects falling within the scope of the Directive and requiring planning permission in England and Wales.

For such projects the Directive was given legal effect through the Town and Country Planning which came into force on 14 March 1999 and apply to relevant planning applications lodged on or after that date. The full text of the Regulations is published separately by The Stationery Office. For ease of reference, those parts of the Regulations which list the types of project to which they apply, and specify what information an environmental statement must contain, are reproduced in Appendices 2 to 4 to this booklet. The booklet is not intended to be an authoritative interpretation of the law and does not remove the need to refer to the Regulations.

Formal guidance on procedures under the Regulations, directed principally at local planning authorities, was issued in DETR Circular 2/99 (Welsh Office Circular 11/99). Although the present booklet, like its predecessor, is meant to be reasonably self-contained, developers may need to refer to that Circular, particularly for fuller information on how planning authorities are expected to judge the significance of a project's likely effects for the purpose of deciding whether environmental impact assessment is required.

Parts 1 and 2 also give some general guidance, applicable to all types of project, on the nature of environmental impact assessment and on the practical aspects of preparing an environmental statement. Part 3 gives a brief account of the procedures which apply to other projects within the scope of the Directive but which are not approved under planning procedures, for example, motorways, harbour works and long distance pipe-lines. It also deals very briefly with environmental impact assessment procedures in Scotland and Northern Ireland. For the detailed requirements, reference will need to be made to the relevant statutory instruments and associated guidance.

Throughout this booklet, the term 'environmental impact assessment' (EIA) is used to describe the whole process whereby information about the environmental effects of a project is collected, assessed and taken into account in reaching a decision on whether the project should go ahead or not. The process was formerly referred to in the UK as 'environmental assessment' (EA). An 'environmental statement' is a document setting out the developer's own assessment of a project's likely environmental effects, which is prepared and submitted by the developer in conjunction with the application for consent. Go to table of contents

PROJECTS WHICH REQUIRE PLANNING PERMISSION

The term 'environmental impact assessment' (EIA) describes a procedure that must be followed for certain types of project before they can be given 'development consent'. The procedure is a means of drawing together, in a systematic way, an assessment of a project's likely significant environmental effects. This helps to ensure that the importance of the predicted effects, and the scope for reducing them, are properly understood by the public and the relevant competent authority before it makes its decision.

Environmental impact assessment enables environmental factors to be given due weight, along with economic or social factors, when planning applications are being considered. It helps to promote a sustainable pattern of physical development and land and property use in cities, towns and the countryside. If properly carried out, it benefits all those involved in the planning process.

From the developer's point of view, the preparation of an environmental statement in parallel with project design provides a useful framework within which environmental considerations and design development can interact. Environmental analysis may indicate ways in which the project can be modified to avoid possible adverse effects, for example through considering more environmentally friendly alternatives. Taking these steps is likely to make the formal planning approval stages run more smoothly.

For the planning authority and other public bodies with environmental responsibilities, environmental impact assessment provides a basis for better decision making. More thorough analysis of the implications of a new project before a planning application is made, and the provision of more comprehensive information with the application, should enable authorities to make swifter decisions. While the responsibility for compiling the environmental statement rests with the developer, it is expected that the developer will consult those with relevant information, and the Regulations specifically require that public authorities which have information in their possession which is relevant to the preparation of the environmental statement should make it available to the developer.

The general public's interest in a major project is often expressed as concern about the possibility of unknown or unforeseen effects. By providing a full analysis of a project's effects, an environmental statement can help to allay fears created by lack of information. At the same time, early engagement with the public when plans are still fluid can enable developers to make adjustments which will help to secure a smoother passage for the proposed development and result in a better environmental outcome.

The environmental statement can also help to inform the public on the substantive issues which the local planning authority will have to consider in reaching a decision. It is a requirement of the Regulations that the environmental statement must include a description of the project and its likely effects together with a summary in non-technical language. One of the aims of a good environmental statement should be to enable readers to understand for themselves how its conclusions have been reached, and to form their own judgements on the significance of the environmental issues raised by the project.

Environmental impact assessment can therefore be helpful to all those concerned with major projects. The following paragraphs describe the procedures for deciding whether EIA is necessary in a particular case and, where it is, for carrying out the assessment. The procedure is intended to make the most of the potential benefits of EIA, while keeping the process as simple and flexible as possible, and avoiding any duplication of existing planning procedures.

WHEN IS ENVIRONMENTAL IMPACT ASSESSMENT NEEDED?

The Regulations apply to two separate lists of projects:

i. 'Schedule 1 projects', for which EIA is required in every case;

ii. 'Schedule 2 projects', for which EIA is required only if the particular project in question is judged likely to give rise to significant environmental effects. Lists of Schedule 1 and Schedule 2 projects are given in, respectively, Appendices 2 and 3 to this booklet.

For Schedule 1 projects, whether or not a particular project falls within the scope of the Regulations will normally be clear: several of the definitions of Schedule 1 projects incorporate an indication of scale, in the form of a quantified threshold, which clearly identifies the projects requiring EIA. Where there is any doubt about a project's inclusion in Schedule 1, the procedures described in below can be used to obtain an opinion from the planning authority or a direction from the Secretary of State (or, in Wales, the National Assembly for Wales).

For the much longer list of Schedule 2 projects, the issue turns on the likelihood of 'significant environmental effects'. For the different types of project described in column 1 of Schedule 2, the 1999Regulations introduced a system of thresholds and criteria, shown in column 2,as a method of discounting development which is not likely to have significant effects on the environment. For development where the applicable threshold or criterion is not exceeded or met, EIA is not normally required. However, even where the threshold or criterion is not met or exceeded, EIA may be required if the proposed development is in, or partly in, a 'sensitive area.' In exceptional circumstances the Secretary of State (or, in Wales, the National Assembly for Wales) may exercise his power under the Regulations to direct that a particular type of Schedule 2 development requires EIA even if it is not to be located in a sensitive area and does not exceed or meet the applicable threshold or criterion.

The more environmentally sensitive the location, the more likely it is that the effects of development will be significant and that EIA will be required. That is why the thresholds and criteria do not apply where development is proposed in, or partly in, a `sensitive area' as defined in the Regulations. Such areas include Sites of Special Scientific Interest (SSSIs), National Parks, Areas of Outstanding Natural Beauty, the Broads, World Heritage Sites and scheduled monuments. There is no general presumption that every Schedule 2 development in a sensitive area will require EIA. Nevertheless, in the case of development to be located in or close to SSSIs, especially those which are also international conservation sites such as Ramsar sites or Special Protection Areas for birds, the likely environmental effects will often be such as to require EIA.

HOW 'SIGNIFICANCE' WILL BE ASSESSED

Developments which meet or exceed the applicable threshold are considered on a caseby-case basis. For the purpose of determining whether EIA is necessary, those of the selection criteria set out in Schedule 3 to the Regulations which are relevant to the proposed development, must be taken into account. The selection criteria fall into the three broad headings: characteristics of the development, location of the development, and characteristics of the potential impact.

For obvious reasons there can be no general definition of what constitutes significance.

General guidance on how to assess `significance' is contained in DETR Circular 2/99 (Welsh Office Circular 11/ 99);and rulings may be obtained from the local planning authority or the Secretary of State (or, in Wales, the National Assembly for Wales) on whether EIA is required in particular cases. Essentially the Circular suggests that there are three main criteria of significance:

i. major developments which are of more than local importance;
ii. developments which are proposed for particularly environmentally sensitive or vulnerable locations;
iii. developments with unusually complex and potentially hazardous environmental effects.

These are very general guidelines and, to assist in their application to particular cases, the Circular also sets out indicative thresholds and criteria by reference to particular categories of development listed in Schedule 2 to the Regulations. These are reproduced in the last column of the Table in Appendix 3 to this booklet. It will be obvious that none of these guidelines can be applied as hard and fast rules; circumstances are bound to vary greatly from case to case. Some large-scale projects which exceed the indicative thresholds may not be significant enough to require EIA; some smaller projects, particularly in sensitive locations, may be candidates for EIA. Nevertheless, the guidance in the Circular should provide a starting point for consideration by the developer and the planning authority of the need for EIA. If the matter is referred to the Secretary of State (or, in Wales, the National Assembly for Wales), he will have regard to the published criteria.

OBTAINING A RULING ON THE NEED FOR EIA

Where there is a possibility that a proposed development will require environmental impact assessment, developers are advised to consult the relevant planning authority well in advance of a planning application. Developers can decide for themselves that a given project falls within the scope of the Regulations so that an environmental statement will be needed. But the Regulations also provide a procedure which enables developers to apply to the planning authority for an opinion ('screening opinion') on whether EIA is needed in a particular case, as soon as a basic minimum of information can be provided about the proposal. This must include a plan on which the site of the proposed development is identified, and a brief description of its nature and purpose and of its possible effects on the environment. This can, of course, be supplemented with other information if the developer wishes.

Where such information can be provided, the developer may approach the planning authority at any time for an opinion on the need for EIA. This can be done well in advance of any formal planning application, although any approach to the planning authority before the planning application stage is entirely voluntary. Where such an approach is made, the planning authority must give its opinion within three weeks, unless the developer agrees to a longer period. The planning authority may request further information from the developer, but this in itself does not extend the three-week time limit, unless the developer agrees. The planning authority must make its determination available for public inspection at the place where the appropriate register (or relevant section of that register) is kept.

Where the planning authority expresses the opinion that a particular proposal requires EIA-whether in response to a request from a developer prior to a planning application, or following a planning application-it must provide a written statement giving clear and precise reasons for its opinion. Both that statement and the developer's application for an opinion are then made available for public inspection at the same place as the register.

A developer who is dissatisfied with the planning authority's opinion that EIA is required may refer the matter to the Secretary of State (or, in Wales, the National Assembly for Wales). The developer is simply required to copy the relevant papers to the Government Office in the region concerned (or the Assembly) and add any representations which are considered to be appropriate in the light of the planning authority's statement.

The Secretary of State (or the Assembly) will then normally give a direction within three weeks of the developer's application; and, if the direction is to the effect that EIA is required, it will be accompanied by a statement of reasons. The developer may, when requesting a screening opinion from the planning authority, simultaneously request an opinion on what should be included in any environmental statement.

The broad intention of this procedure is to ensure that developers can obtain a clear ruling on the need for EIA well before they reach the stage of lodging a formal planning application. This should minimise the possibility of delay or uncertainty at that stage. Where the matter is not raised until a formal planning application is lodged, the developer risks serious delay if either the planning authority or the Secretary of State (or, in Wales, the National Assembly for Wales) rules that an environmental statement must be prepared. No action will be taken on the planning application until the developer has prepared an environmental statement and submitted it to the planning authority.

In most cases this procedure will give developers a firm decision on the need for EIA as soon as they can provide basic information about their project. There may occasionally be cases where the receipt of information directly leads the Secretary of State (or, in Wales, the National Assembly for Wales) to issue a direction when there has been no request for one from the developer; or to overrule a planning authority's opinion that EIA is not required; or even, very exceptionally, to reverse an earlier direction. This could, for instance, happen where a development proposal came to the Secretary of State's (or the Assembly's) notice on 'call-in'; or where third party representations drew attention to aspects of a proposed development which were not known to the Secretary of State (or the Assembly) when the initial direction was given. Exceptionally, also, a planning authority which has given a pre-application opinion that EIA is not required might consider it necessary to reverse that decision when the planning application is formally submitted. In that case the developer could, of course, apply to the Secretary of State (or the Assembly) for a direction.

'PERMITTED DEVELOPMENT RIGHTS' (PDRS)

Developments which do not require planning permission because of the provisions of the Town and Country Planning (General Permitted Development) PDRs provided that they do not fall into Schedule 1 or 2 of the EIA Regulations. For developments that do fall within Schedule 1 or 2, the general position is as follows. Schedule 1projects are not permitted development, and always require the submission of a planning application and an environmental statement.

PDRs for Schedule 2 projects which either exceed or meet the applicable threshold or criterion, or are wholly or partly in a sensitive area, are also withdrawn, unless the local planning authority has adopted a screening opinion (or the Secretary of State (or, in Wales, the National Assembly for Wales) has directed) to the effect that EIA is not required. There are exceptions to these provisions in the case of the following classes in Schedule 2 to the 1995 Order: Part 7, Class D of Part 8, Part 11, Class B of Part 12, Class F (a) of Part 17, Class A of Part 20, Class B of Part 20, and Class B of Part

These exceptions exist for a variety of reasons. For example, some relate to projects subject to alternative consent procedures, and others to projects begun before Directive 85/337/EEC came into operation. Further information about the exceptions, and about permitted development and EIA generally, can be found in DETR Circular 2/99 (Welsh Office Circular 11/99).

SIMPLIFIED PLANNING ZONES AND ENTERPRISE ZONES

Special considerations apply to projects proposed for simplified planning zones (SPZs) and enterprise zones (EZs).

Simplified Planning Zones

All Schedule 1 projects are excluded from the scope of SPZs and therefore require EIA as part of the application for planning permission. Where the terms of SPZ schemes would permit Schedule 2 projects to be undertaken without specific planning permission, developers must notify the planning authority that they intend to undertake such development.

This notification will give the planning authority and, where appropriate, the Secretary of State (or, in Wales, the National Assembly for Wales) the opportunity to consider whether the project is likely to give rise to significant environmental effects. If so, planning permission will be required in the normal way and an environmental statement must be prepared. If not, the project will enjoy the benefit of the general permission granted by the SPZ scheme, and no separate application for planning permission will be necessary.

Enterprise Zones

If the planning scheme allows for Schedule 1 projects, EIA must be carried out. As regards Schedule 2 projects, the same procedures apply as those for SPZs described above.

The arrangements in SPZs and EZs are explained in DETR Circular 2/99 (Welsh Office Circular 11/99).

Go to table of contents

Preparing an Environmental Statement: the Planning Procedures

- Preliminary consultations
- Content of the environmental statement
- Statutory and other consultees; the general public
- Techniques of assessment; sources of advice
- Submission of environmental statement with planning application
- Handling by the planning authority
- Requests for further information
- Determination of application
- Appeals and call-ins
- Procedural stages

The developer is responsible for preparing the environmental statement which must be submitted with the planning application. A developer may choose to engage consultants for some or all of the work. The preparation of the statement should be a collaborative exercise involving discussions with the local planning authority, statutory consultees and possibly other bodies as well. There is no prescribed form of statement, provided that the requirements of the Regulations are met. The Regulations enable a developer, before making a planning application, to ask the local planning authority for its formal opinion ('scoping opinion') on the information to be included in an environmental statement.

The developer can therefore be clear as to what the authority considers are the aspects of the environment which would be affected. The request may be made at the same time as for the screening opinion, and must be accompanied by the same information as for the screening opinion. A developer may also wish to submit a draft outline of the environmental statement, indicating what the developer thinks the main issues are, as a focus for the authority's consideration. The authority must consult certain bodies and the developer before adopting a scoping opinion, and must adopt the scoping opinion within five weeks of receiving the request. This period may be extended provided that the developer agrees. The scoping opinion must be kept available for public inspection for two years, with the request (including documents submitted by the developer as part of that request), at the place where the planning register is kept.

There is no provision for an appeal to the Secretary of State (or, in Wales, the National Assembly for Wales) if the developer and local planning authority disagree about the content of an environmental statement. If an authority fails to adopt a scoping opinion within five weeks (or any agreed extension), the developer may apply to the Secretary of State (or the Assembly) for a scoping direction.

The application must be accompanied by the previous documents relating to the request for a scoping opinion, plus any additional representations the developer wishes to make. The developer should also send a copy of the request and any representations to the authority. The Secretary of State (or the Assembly) must make a scoping direction within five weeks of receiving the request, or such longer period as he may reasonably require. He must consult the developer and certain bodies beforehand. A copy of the scoping direction will be sent to the developer and the authority.

The aim should be to provide as systematic and objective an account as is possible of the significant environmental effects to which the project is likely to give rise. Where the statement embodies or summarises the conclusions of more detailed work, sufficient information should be provided to enable those who wish to do so to verify the statement's conclusions and to identify the source of the information provided. The environmental statement must contain a non-technical summary which will enable non-experts to understand its findings.

Preliminary Consultations

One of the main emphases of the process of environmental impact assessment is on the need for fuller and earlier consultation by the developer with bodies which have an interest in the likely environmental effects of the development proposal. If important issues are not considered at a very early stage, they may well emerge when a project's design is well advanced, and necessitate rethinking and delay. Ideally, EIA should start at the stage of site selection and (where relevant) process selection, so that the environmental merits of practicable alternatives can be properly considered.

While a developer is under no formal obligation to consult about the proposal before the submission of a formal planning application, there are good practical reasons for doing so. Authorities will often possess useful local and specialised information which is relevant to a project's design, and officers may be able to give preliminary advice about local problems and about those aspects of the proposal that are likely to be of particular concern to the authority.

The timing of such informal consultations is at the developer's discretion; but it will generally be advantageous for them to take place as soon as the developer is in a position to provide sufficient information about the proposal to form a basis for discussion. The developer can ask that any information provided at this preliminary stage should be treated in confidence by the planning authority and any other consultees. If the developer is seeking a formal opinion from the planning authority on the need for environmental impact assessment, the information about the project which accompanies that request will be made public by the authority.

Content of the Environmental Statement

Developers and authorities should discuss the scope of an environmental statement before its preparation is begun.

The formal requirements as to the content of environmental statements are set out in Schedule 4 to the Regulations, which is reproduced in Appendix 4 to this booklet. The information is given under separate headings: Part I and Part II. The statement must include at least the information included in Part II, and such of the information in Part I as is reasonably required to assess the environmental effects of the development and which the applicant can reasonably be required to complete. As a practical guide to the range of issues which may need to be considered, developers may find it helpful to use the checklist at Appendix 5 to this booklet as a basis for their discussions with the planning authority. The checklist is not meant to be regarded as a prescribed framework for all environmental statements. Its main purpose is to act as a guide or agenda for the preliminary discussions about the scope of the statement.

The comprehensive nature of the checklist at Appendix 5 should not be taken to imply that all environmental statements should cover every conceivable aspect of a project's potential environmental effects at the same level of detail. They should be tailored to the nature of the project and its likely effects. Whilst every environmental statement should provide a full factual description of the project, the emphasis of Schedule 4 is on the main or significant effects to which a project is likely to give rise.

In some cases, only a few of the aspects set out in the checklist will be significant in this sense and will need to be discussed in the statement in any great depth. Other issues may be of little or no significance for the particular project in question, and will need only very brief treatment, to indicate that their possible relevance has been considered.

It should be noted that developers are now required to include in the environmental statement an outline of the main alternative approaches to the proposed development that they may have considered, and the main reasons for their choice. It is widely regarded as good practice to consider alternatives, as it results in a more robust application for planning permission. Also, the nature of certain developments and their location may make the consideration of alternatives a material consideration. Where alternatives are considered, the main ones must be outlined in the environmental statement.

Even where a local planning authority has adopted a scoping opinion, the developer is responsible for the content of the statement which is finally submitted. Developers should bear in mind that planning authorities have powers to call for additional information when considering environmental statements and planning applications, and that they are likely to use those powers if they consider that aspects of a submitted environmental statement are inadequate. There is no provision for any disagreement between the developer and the planning authority over the content of an environmental statement to be referred to the Secretary of State (or, in Wales, the National Assembly for Wales), except through normal planning appeal procedures.

Developers should consider at an early stage whether an assessment of environmental effects may also be required under another European Community Directive, such as the Habitats Directive (92/43/EEC), the Wild Birds Directive(79/409/EEC), the Integrated Pollution Prevention and Control Directive(96/61/EC) or the Control of Major Accident Hazards Directive (96/82/EC).Although the requirements of these and of the EIA Directive are all independent of each other, there are clearly links between them. Where more than one regime applies, developers could save unnecessary time and effort if they identify and co-ordinate the different assessments required. Advice on the links between the EIA system and the requirements of the Habitats Regulations is offered in PPG 9on Nature Conservation (or, in Wales, Planning Guidance (Wales) Planning Policy First Revision), and on the links between the Town and Country Planning system and the IPPC authorisation system in PPG 23 on Planning and Pollution Control (or, in Wales, Planning Guidance (Wales) Planning Policy First Revision and Planning Guidance (Wales) Technical Advice Note (Wales) 5 'Nature Conservation and Planning').

STATUTORY AND OTHER CONSULTEES; THE GENERAL PUBLIC

The Regulations give a particular role in environmental impact assessment to those public bodies with statutory environmental responsibilities who must be consulted by the planning authority before a Schedule I or a Schedule 2 planning application is determined. A full list of these statutory consultation bodies is given in Appendix 6 to this booklet.

Where the planning authority or the Secretary of State(or, in Wales, the National Assembly for Wales) rules that EIA is required, those bodies which are statutory consultees for the particular project in question will be notified and the developer will be informed accordingly.

The effect of this notification is to put those bodies under an obligation to provide the developer (on request) with any information in their possession which is likely to be relevant to the preparation of the environmental statement. An example might be information held by English Nature (or, in Wales, the Countryside Council for Wales) about the ecology of a particular area, which could be relevant to the assessment of a project's likely effects.

It is up to the developer to approach the statutory consultees and indicate what sort of information would be helpful in preparing the environmental statement. The obligation on statutory consultees relates only to information already in their possession; they are not required to undertake research on behalf of the developer. Nor, at this stage, would consultees be expected to express a view about the merits of the proposal; their views on merits are invited at a later stage. Consultees may make a reasonable charge to cover the cost of making information requested by a developer available.

Developers should also consider whether to consult the general public, and non-statutory bodies concerned with environmental issues, during the preparation of the environmental statement. Bodies of this kind may have particular knowledge and expertise to offer. Some are national organisations, for instance, the Royal Society for the Protection of Birds; inmost areas there are also active local amenity societies and environmental groups.

While developers are under no obligation to publicise their proposals before submitting a planning application, consultation with local amenity groups and with the general public can be useful in identifying key environmental issues, and may put the developer in a better position to modify the project in ways which would mitigate adverse effects and recognise local environmental concerns. It will also give the developer an early indication of the issues which are likely to be important at the formal application stage if, for instance, the proposal goes to public inquiry.

TECHNIQUES OF ASSESSMENT; SOURCES OF ADVICE

Extensive literature is available on how to assess the effects on the environment of particular processes and activities. The assessment techniques used, and the degree of detail in which any particular subject is treated in an environmental statement, will depend on the character of the proposal, the environment which it is likely to affect, and the information available. While a careful study of the proposed location will generally be needed (including environmental survey information), original scientific research will not normally be necessary. The local planning authority and statutory consultees may be able to advise the developer on sources of specialist information, for example, about particular local conditions.

Environmental statements will often need to recognise that there is some uncertainty attached to the prediction of environmental effects. Where there is uncertainty, it needs to be explicitly recognised. Uncertainty is not in itself a reason for discounting the importance of particular potential environmental effects, simply because other effects can be more confidently predicted.

Submission of Environmental Statement with Planning Application

To enable a planning application to be processed as quickly as possible, it is in the developer's interest to submit an environmental statement at the same time as the application is made. It will be for the planning authority to judge how much information is required in the particular case, but the preparation of an environmental statement is bound to require the developer to work out proposals in some detail; otherwise any thorough appraisal of likely effects will be impossible. Where an application is in outline, the planning authority will still need to have sufficient information on a project's likely effects to enable it to judge whether the development should take place or not. The information given in the environmental statement will have an important bearing on whether matters may be reserved in an outline permission; it will be important to ensure that the development does not take place in a form which would lead to significantly different effects from those considered at the planning application stage.

When the developer submits an environmental statement at the same time as the planning application, three further copies must also be submitted for onward transmission by the planning authority to the Secretary of State (or, in Wales, the National Assembly for Wales). The developer is also required to provide the planning authority with sufficient copies of the environmental statement to enable one to be sent to each of the statutory consultees. Alternatively, the developer may send copies of the statement directly to the consultees. When submitting the application, the developer must inform the planning authority of the name of every body-whether or not it is a statutory consultee-to which a copy of the statement has been sent.

The developer should make a reasonable number of copies of the statement available for members of the public. A reasonable charge reflecting printing and distribution costs may be made.

HANDLING BY THE PLANNING AUTHORITY

Where the environmental statement is submitted with the planning application, the local planning authority will arrange for a notice to be published in a local newspaper and displayed at or near the site of the proposed development. If the environmental statement is submitted after the planning application, responsibility for publicising it falls to the developer.

The planning authority will place the planning application on Part I of the planning register, together with the environmental statement. The authority and the developer may wish to consider the need for further publicity at this stage, for example, publication of further details of the project in a local newspaper, or an exhibition. The planning authority will also need to notify statutory consultees of the application (unless the developer has already done so) and invite them to comment on the environmental statement. Consultees must be allowed at least 14 days from receipt of the statement in which to comment before a decision is taken. It will often be useful for the planning authority to discuss the project with consultees who have a particular interest in its environmental effects before reaching its conclusions on the planning applications.

The copies of the environmental statement that are forwarded by the planning authority to the Secretary of State (or, in Wales, the National Assembly for Wales), will assist in monitoring. In those exceptional circumstances where a proposed development is likely to have significant effects on the environment in another country, they will also enable certain international obligations for exchange of information with other countries to be met.

Requests for Further Information

Where the planning authority considers that the information provided in the developer's environmental statement, together with that available to the authority from other sources, is insufficient to permit a proper evaluation of the project's likely environmental effects, the authority can require further information, or evidence to verify the information that has already been provided.

The use of these powers should not normally be necessary, especially if the parties have worked together during the preparation of the environmental statement. Nevertheless, further consultation between the planning authority and the developer may be necessary at this stage, in particular to consider comments made by consultees and, possibly, amendments to the proposal to meet objections that have been raised. Where an authority considers that it does not have the necessary expertise to evaluate the information contained in an environmental statement, it may decide to seek advice from consultants or other suitably qualified persons or organisations.

Determination of Application

The planning authority is required to determine a planning application which is the subject of environmental impact assessment within 16 weeks from the date of receipt of the environmental statement, unless the developer agrees to a longer period. In determining the application, the authority is, of course, required to have regard to the environmental statement, as well as to other material considerations.

As with any other planning application, the planning authority may refuse permission or grant it with or without conditions.

The planning authority cannot take the view that a planning application is invalid because it considers that an inadequate environmental statement has been submitted or because the developer has failed to provide any further information required under the powers described above. However, if the developer fails to provide enough information to complete the environmental statement, the application can be determined only by refusal.

Appeals and Call-ins

The right of appeal to the Secretary of State (or, in Wales, the National Assembly for Wales) against an adverse decision by a planning authority (or against an authority's failure to determine an application within the 16-week time limit) is the same for planning applications to which the EIA Regulations apply as for other applications. Similarly, the Secretary of State's (or the Assembly's) power to call in a planning application applies in these cases.

Where an environmental statement has been prepared to accompany a planning application, the information which it contains will be among the material considerations which an Inspector will take into account in considering an appeal. The Secretary of State (or the Assembly) and Inspectors, like the planning authority, have power to request the developer to provide further information where they consider that the environmental statement is inadequate as it stands.

Any additional information provided by the developer in response to such a request will be made available to all parties to an appeal.

Procedural Stages

Appendix 7 to this booklet provides illustrative flowcharts for the five main procedural stages: application to the planning authority for a screening opinion; application to the Secretary of State (or, in Wales, the National Assembly for Wales) for a screening direction; application to the planning authority for a scoping opinion; application to the Secretary of State (or the Assembly) for a scoping direction; and submission of an environmental statement to the planning authority in conjunction with a planning application.

Arrangements for other Projects

- The trunk road network
- Oil and gas pipe-lines
- Offshore oil and gas projects
- Nuclear power stations
- Other power stations and overhead power lines
- Forestry projects
- Land drainage improvements
- Ports and harbours
- Marine fish farming
- Marine dredging for minerals
- Transport and Works Act 1992
- Projects arising in Scotland and Northern Ireland

However, those Parts specifically describe the arrangements for EIA of projects in England and Wales which are approved through the planning system. Part 3 gives brief guidance on the legislative provisions on EIA for projects insofar as they are outside the scope of the planning system.

For all the following projects the Regulations described, which implement the Directive, include provisions for seeking a scoping opinion from the competent authority, for consulting statutory and other bodies, for the submission of an environmental statement in the prescribed form, and for publicising the statement. Where applicable, the Regulations disapply any thresholds for Annex II projects which are at least partly in a 'sensitive area', this term being defined in the same or almost the same way as in the planning EIA Regulations.

PROJECTS NOT SUBJECT TO PLANNING CONTROL

The Trunk Road Network

EIA is mandatory for projects to construct new motorways and certain other roads, including those with four or more lanes, and for certain road improvements.

Roads in England and Wales for which the Secretary of State for Transport or the National Assembly for Wales is the highway authority, are approved under procedures set out in the Highways Act 1980. These roads-the trunk road network-currently include nearly all motorways. The Department concerned will normally consult the public widely about alternative routes before selecting a preferred route for a new road. Once the preferred route has been announced, detailed design work is carried out leading to the publication of statutory orders. In most cases these orders are the subject of a public inquiry held by an independent inspector.

For these roads, the Directive has been implemented in England and Wales by the Highways (Assessment of Environmental Effects) Regulations 1999 (SI No 369) http://www.legislation.hmso.gov.uk/si/si1999/19990369.htm. These Regulations replace the Highways (Assessment of Environmental Effects) Regulations 1988 (SI No 1241), except for projects begun before the new Regulations came into force on 13 March 1999.

The Regulations incorporate an exclusive threshold for the purpose of screening Annex II projects: only those in which the area of the completed works and the area occupied during the road construction or improvement period exceeds one hectare, or are wholly or partly in a sensitive area, need to be screened for EIA.

Roads developed by local planning authorities, and roads developed by private developers, require planning permission and the provisions described in Parts I and 2 of this booklet apply to such roads.

Oil and Gas Pipe-lines

Oil and gas pipe-lines with a diameter of more than 800 millimetres and a length of more than 40 kilometres fall within Annex I of the Directive. Pipe-line installations which are below either of these thresholds fall within Annex II. Oil and gas pipe-lines 10 miles long or less are approved under planning legislation.

Onshore Pipe-line Works

Proposed onshore pipes (but not those of public gas transporters-see below) which are more than 10 miles long requirc a pipe-line construction authorisation (PCA) from the Secretary of State for Trade and Industry under the Pipe-lines Act 1962. The Pipe-line Works (Environmental Impact Assessment) Regulations 2000 (SI No 1928), which replace the provisions relating to pipe-lines in the Electricity and Pipe-line Works (Assessment of Environmental Effects) Regulations 1990, implement the Directive in England, Wales and Scotland in respect of pipe-lines which require a PCA. They require 'relevant pipe-line works' to be subject to EIA unless the Secretary of State makes a determination that EIA is not necessary. Public gas transporter pipe-line works

Under the licensing regime introduced by the Gas Act 1995, companies wishing to convey gas may be licensed as public gas transporters. As such, they are exempt from the need to obtain authorisation for their pipes under the Pipe-lines Act 1962.

The Public Gas Transporter Pipe-line Works (Environmental Impact Assessment) Regulations 1999 came into force on 15 July 1999 and implement the Directive in England, Wales and Scotland. They require a public gas transporter proposing to undertake pipe-line works which fall in Annex I of the Directive to submit an environmental statement and apply to the Secretary of State for Trade and Industry for consent to carry them out. Pipe-line works in Annex II of the Directive may be subject to EIA if they have a design operating pressure exceeding 7 bar gauge or either wholly or in part cross a sensitive area. In these circumstances, the public gas transporter must, before commencing construction, either obtain a determination from the Secretary of State that an environmental statement is not required, or give notice that it intends in any event to produce an environmental statement. The Regulations also provide for the Secretary of State to require an environmental statement where proposed works do not meet these criteria but nevertheless it is considered that there are likely to be significant environmental effects.

Offshore Oil and Gas Projects

The Directive covers the extraction of petroleum and natural gas for commercial purposes where the amount extracted exceeds 500 tonnes per day in the case of petroleum and 500 000 cubic metres per day in the case of gas.

For such projects, the Offshore Petroleum Production and Pipe-lines (Assessment of Environmental Effects) Regulations 1999 (SI No 360) implement the Directive. They apply to the whole of the UK. They replace the Offshore Petroleum and Pipe-lines (Assessment of Environmental Effects) Regulations 1998 (SI No 968), which are revoked except for saving provisions for applications for consent received before 14 March 1999.

The Regulations are linked to licences granted by the Secretary of State for Trade and Industry under the

Petroleum Act 1998. Such licences require the Secretary of State's consent to be obtained to the drilling of a well, the getting of petroleum (where the amount exceeds 500 tonnes per day in the case of oil and 500 000 cubic metres per day in the case of gas) and the erection of any structure in connection with a development.

They require, subject to an exception for floating installations whose use commenced before 30 April 1998 (when the 1998 Regulations came into force), the Secretary of State's consent for the use of a floating installation in prescribed circumstances. The Regulations also require the Secretary of State's consent to the use of a mobile installation for the purposes of the extraction of petroleum where the main purpose of such extraction is the testing of a well.

For projects which fall outside Annex I of the Directive, the Secretary of State decides whether the project is likely to have significant effects and therefore whether EIA is required.

Nuclear Power Stations

Annex 1.2 of the Directive covers nuclear power stations and other nuclear reactors, and now includes their dismantling or decommissioning, so EIA is mandatory for these projects. It does not cover research installations for the production and conversion of fissionable and fertile material, whose maximum power does not exceed 1 kilowatt continuous thermal load.

For new nuclear power stations over 50 megawatts, the Secretary of State for Trade and Industry is the competent authority, and the Directive has been implemented in England and Wales by the Electricity Works (Assessment of Environmental Effects) Regulations 2000 (SI No 1927), which also applies to thermal power stations and overhead power lines. Power stations of less than 50 megawatts, and other buildings to house small nuclear reactors, are given consent through the normal planning system and are subject to the provisions described in Parts 1 and 2 of this booklet.

The Directive's provisions on the dismantling and decommissioning of nuclear power stations and reactors have been implemented by the Nuclear Reactors (Environmental Impact Assessment for Decommissioning) Regulations 1999 (SI No 2892) which apply to England, Wales and Scotland. They prohibit dismantling or decommissioning from being carried out without the consent of the Health and Safety Executive.

A licensee who applies for consent is required to provide the Executive with an environmental statement. The Regulations also apply to changes to existing dismantling or decommissioning projects which may have significant effects on the environment (Annex II.13 of the Directive).

Other Power Stations and Overhead Power Lines

Annex I of the Directive includes thermal power stations with a heat output of 300 megawatts or more, and the construction of overhead power lines with a voltage of 220 kilovolts or more and a length of more than 15 kilometres.

The construction or extension of power stations exceeding 50 megawatts, and the installation of overhead power lines, require consent from the Secretary of State for Trade and Industry under sections 36 and 37 of the Electricity Act 1989, and the requirement for EIA has been implemented as part of the procedure for applications under those provisions. The Electricity Works (Environmental Impact Assessment) Regulations 2000 (SI No 1927) require EIA for all projects which fall within Annex I of the Directive. They also require proposed power stations not covered by Annex I, and all overhead power lines of 132 kilovolts or more, to be screened for EIA.

Overhead power lines below 132 kilovolts are not normally expected to have significant environmental effects and do not need to be considered for EIA unless local circumstances demand it. The Regulations also make provisions relating to nuclear power stations.

The Regulations apply to England and Wales and replace that part of the Electricity and Pipe-line Works (Assessment of Environmental Effects) Regulations 1990 dealing with power stations and overhead cables. The 1990 Regulations, however, remain in force in respect of applications lodged before the new Regulations came into force. Power stations of 50 megawatts or less are approved under planning legislation and are subject to the provisions described in Parts 1 and 2 of this booklet.

LAND DRAINAGE IMPROVEMENTS

Land drainage projects fall within Annex II. 1(c) of the Directive. New land drainage works, including flood defence works and defences against the sea, require planning permission and the provisions described in Parts 1 and 2 of this booklet apply to such works. Land drainage improvement works undertaken by drainage bodies are permitted development under the Town and Country Planning (General Permitted Development) Order 1995 and are therefore exempt from planning permission. As such works might have significant effects on the environment, the principles of EIA need to be applied to them. This is done through the Environmental Impact Assessment (Land Drainage Improvement Works) Regulations 1999 (SI No 1783), which apply to England and Wales and revoke and replace earlier Regulations which came into force in 1988 and were amended in 1995.

The 1999 Regulations require a drainage body to consider whether proposed improvement works are likely to have significant effects on the environment. Where the drainage body considers that there are unlikely to be such effects, it must publicise its intention to carry out the works. If the drainage body receives representations that there are likely to be significant effects but it still thinks otherwise, it must apply for a determination to the appropriate authority (Secretary of State for the Environment, Food and Rural Affairs or, in Wales, the National Assembly for Wales).

Where the drainage body concludes that the works are likely to have significant environmental effects, it must publicise its intention to prepare an environmental statement and notify specified consultation bodies (English Nature, the Countryside Agency and any other authority or organisation which might have an interest).

The environmental effects of the improvement works must be assessed in the light, in particular, of the environmental statement and any representations received. If there are no objections, the drainage body may determine that it will proceed with the works. If there are objections, the proposal must be referred to the appropriate authority for a determination giving or refusing consent to the works. The determination must be publicised.

Ports and Harbours

Annex 1.8 of the Directive includes trading ports, piers for loading and unloading connected to land and outside ports (excluding ferry piers) which can take vessels of over 1350 tonnes. Annex II.10(e) includes construction of harbours and port installations, including fishing harbours, which do not fall within Annex I.

Proposed development at ports and harbours down to the low water mark is subject to Town and Country Planning EIA Regulations. For other works, the requirements of the Directive have been implemented in England, Wales and Scotland by the Harbour Works (Environmental Impact Assessment) Regulations 1999 (SI No 3445).

These Regulations incorporate a revised Schedule 3 to the Harbour Works Act 1964, which gives the procedures for harbour revision orders and harbour empowerment orders, and replace the Harbour Works (Assessment of Environmental Effects) (No 2) Regulations 1989 governing harbour works not needing to be authorised by harbour order.

Whether or not proposed works are subject to a harbour order, the appropriate Minister (generally the Secretary of State for Transport) is responsible for deciding whether EIA is required and, if so, whether the works should proceed in the light of the assessment and the comments of environmental bodies.

In general, works under harbour empowerment orders are likely to fall within Annex I of the Directive and require EIA in every case, whereas works under harbour revision orders are likely to fall within Annex II and the need for EIA will depend on whether there are likely to be significant environmental effects. Proposed works which are subject to harbour orders are not subject to EIA if the area of works comprised in the project does not exceed 1 hectare, unless any part falls within a `sensitive area' or the Secretary of State decides that EIA is needed in a particular case.

Marine Fish Farming

Intensive fish farming is an Annex II project. Developments of onshore fish farming facilities may require planning permission and the provisions described in Parts 1 and 2 of this booklet apply to proposals for such facilities.

Off-shore facilities do not require planning permission but require a lease from the Crown Estate Commissioners (or, where appropriate, from the Shetlands Islands Council or Orkney Islands Council). The Environmental Impact Assessment (Fish Farming in Marine Waters) Regulations 1999 (SI No 367), which apply to England, Wales and Scotland, set indicative criteria which, when triggered, require the Commissioners to screen applications to determine whether EIA is needed.

They apply where the development is designed to hold a biomass of 100 tonnes or more, or where it will extend to 0.1 hectare or more of the surface area of the marine waters (including any proposed structures or excavations), or where any part of the development is to be carried out in a sensitive area. The Regulations prohibit a lease from being granted before the environmental statement provided by the developer, and comments on it from consultees and the general public, have been considered.

The Crown Estate Commissioners have, in consultation with the Scottish Executive Rural Affairs Department, issued an Environmental Assessment Guidance Manual for Marine Salmon Farmers which provides guidance to developers on how to conduct EIA and prepare an environmental statement. Guidance on the practical implications of the 1999 Regulations has also been issued by the Scottish Executive Rural Affairs Department.

Marine Dredging for Minerals

Extraction of minerals by marine dredging falls within Annex II.2 (c) of the Directive. The Environmental Impact Assessment and Habitats (Extraction of Minerals by Marine Dredging) Regulations, which are likely to come into force early in 2001, will implement the Directive in waters around the UK (but excluding waters around Scotland). They will also implement the Habitats Directive (92/43/EEC).

Unless the regulator (Secretary of State in respect of English Waters or National Assembly in respect of Welsh waters) has provided, in response to a request, a written determination that proposed dredging is not likely to have significant effects, application for permission must be obtained and the application must be accompanied by an environmental statement. The only exceptions to this are where the Secretary of State has determined that the dredging would form part of a project serving national defence purposes, or where the regulator has determined that it is exempted under Article 2.3 of the Directive.

Transport and Works Act 1992

The Transport and Works Act 1992 (TWA) enables orders to be made authorising the construction or operation of railways, tramways, other guided transport systems and inland waterways; and works interfering with rights of navigation. Orders may also authorise ancillary matters such as compulsory purchase and creating or extinguishing rights over land. Applicants for TWA orders may apply at the same time for a direction that planning permission is deemed to be granted.

The procedures for making applications for orders under Part 1 of the Act are contained in the Transport and Works (Applications and Objections Procedure) (England and Wales) Rules 2000 (SI No 2190), which also implement the Directive. These Rules apply in England and Wales and replace the earlier 1992 Rules (SI No 2902) except for applications made and not determined before the new Rules come into force on 16 October 2000. In England, most applications for orders are determined by the First Secretary of State. Applications for orders relating wholly to Wales fall to be determined by the National Assembly for Wales.

The procedures include provision for opposed orders to be considered by way of a public inquiry, hearing or exchanges of written representations. Schemes that are identified by the Secretary of State or the Assembly as being of national significance are also subject to Parliamentary approval prior to a public inquiry being held.

PROJECTS ARISING IN SCOTLAND AND NORTHERN IRELAND

The arrangements for environmental impact assessment in Scotland and Northern Ireland are broadly similar to those applying in England and Wales, but are mainly subject to separate legislative provisions. The Environmental Impact Assessment (Scotland) Regulations 1999 (SSI 1999/1) implement the Directive in Scotland for projects which are subject to planning permission and for certain trunk road projects and land drainage works. Scottish Executive Development Department Circular 15/1999 explains the Regulations, and Planning Advice Note (PAN 58) on Environmental Impact Assessment also provides background information and guidance on best practice. The preceding section, of this booklet indicates which of those Regulations relating to projects outside the planning system in England and Wales, also apply in Scotland.

The Planning (Environmental Impact Assessment) Regulations (Northern Ireland) 1999 (SR 1999/73) implement the Directive in Northern Ireland for projects subject to planning permission. Development Control Advice Note 10 (Revised 1999) issued by the Planning Service, an agency of the Department of the Environment for Northern Ireland, explains the Regulations.

Appendix 8 includes a list of all implementing legislation currently in force in England and Wales, Scotland and Northern Ireland, and a list of available guidance. Go to table of contents

PRIVATE PROJECTS ON THE ENVIRONMENT

As amended by Council Directive 97/11/EC (adopted by Council 3 March 1997). (Text of Directive 97/I1/EC is available in the Official Journal of the European Communities, 14 March 1997). Text in italics is a change or an addition made by 97/11/EC.

The Council of the European Union Having regard to the Treaty establishing the European Community, and in particular Article 130s (1) thereof, Having regard to the proposal from the Commission 1, Having regard to the opinion of the Economic and Social Committee 2, Having regard to the opinion of the Committee of the Regions 3, Acting in accordance with the procedure laid down in Article 189c of the Treaty 4.

1. Whereas Council Directive 85/337/EEC of 27 June 1985 on the assessment of the effects of certain public and private projects on the environment 5 aims at providing the competent authorities with relevant information to enable them to take a decision on a specific project in full knowledge of the project's likely significant impact on the environment; where as the assessment procedure is a fundamental instrument of environmental policy as defined in Article 130r of the Treaty and of the Fifth Community Programme of policy and action in relation to the environment and sustainable development;
2. Whereas, pursuant to Article 130r (2) of the Treaty, Community policy on the environment is based on the precautionary principle and on the principle(sic) that preventive action should be taken, that environmental damage should as a priority be rectified at source and that the polluter should pay;
3. Whereas the main principles of the assessment of environmental effects should be harmonised and whereas the Member States may lay down stricter rules to protect the environment;
4. Whereas experience acquired in environmental impact assessment, as recorded in the report on the implementation of Directive 85/337/EEC, adopted by the Commission on 2 April 1993, shows that it is necessary to introduce provisions designed to clarify, supplement and improve the rules on the assessment procedure, in order to ensure that the Directive is applied in an increasingly harmonised and efficient manner;

5. Whereas projects for which an assessment is required should be subject to a requirement for development consent; whereas the assessment should be carried out before such a consent is granted;
6. Whereas it is appropriate to make additions to the list of projects which have significant effects on the environment and which must on that account as a rule be made subject to systematic assessment;
7. Whereas projects of other types may not have significant effects on the environment in every case; whereas these projects should be assessed where Member States consider they are likely to have significant effects on the environment;
8. Whereas Member States may set thresholds or criteria for the purpose of determining which such projects should be subject to assessment on the basis of the significance of their environmental effects; whereas Member States should not be required to examine projects below those thresholds or outside those criteria on a case-by-case basis;
9. Whereas when setting such thresholds or criteria or examining projects on a case-by-case basis for the purpose of determining which projects should be subject to assessment on the basis of their significant environmental effects Member States should take account of the relevant selection criteria set out in this Directive; whereas, in accordance with the subsidiarity principle, the Member States are in the best position to apply these criteria in specific instances;
10. Whereas the existence of a location criterion referring to special protection areas designated by Member States pursuant to Council Directive79/409/ EEC of 2 May 1979 on the conservation of wild birds 6 and 92/43/EEC of 21 May 1992 on the conservation of natural habitats and of wild fauna and flora 7 does

not imply necessarily that projects in those areas are to be automatically subject to an assessment under this Directive;

11. Whereas it is appropriate to introduce a procedure in order to enable the developer to obtain an opinion from the competent authorities on the content and the extent of the information to be elaborated and supplied for the assessment; whereas Member States, in the framework of this procedure, may require the developer to provide, inter alia, alternatives for the projects for which it intends to submit an application;
12. Whereas it is desirable to strengthen the provisions concerning environmental impact assessment in a transboundary context to take account of developments a tinternational level;
13. Whereas the Community signed the Convention on Environmental Impact Assessment in a Transboundary Context on 25 February 1991. (Preamble of 85/337/EEC) The Council of the European Communities

Having regard to the Treaty establishing the European Economic Community, and in particular Articles 100 and 235 thereof, Having regard to the proposal from the Commission 8, Having regard to the opinion of the European Parliament 9, Having regard to the opinion of the Economic and Social Committee 10, Whereas the 1973 11 and 1977 12 action programmes of the European Communities on the environment, as well as the 1983 13 action programme, the main outlines of which have been approved by the Council of the European Communities and there representatives of the Governments of the Member States, stress that the best environmental policy consists in preventing the creation of pollution or nuisances at source, rather than subsequently trying to counteract their effects; whereas they affirm the need to take effects on the environment into account at the earliest possible stage in all the technical planning and decision-making processes; whereas to that end, they provide for the implementation of procedures to evaluate such effects; Whereas the disparities between the laws in force in the various Member States with regard to the assessment of the environmental effects of public and private projects may create unfavourable competitive conditions and thereby directly affect the functioning of the common market; whereas, therefore, it is necessary to approximate national laws in this field pursuant to Article 100 of the Treaty; Whereas, in addition, it is necessary to achieve one of the Community's objectives in the sphere of the protection of the environment and the quality of life.

Whereas, since the Treaty has not provided the powers required for this end, recourse should be had to Article 235 of the Treaty; Whereas general principles for the assessment of environmental effects should be introduced with a view to supplementing and coordinating development consent procedures governing public and private projects likely to have a major effect on the environment.

Whereas development consent for public and private projects which are likely to have significant effects on the environment should be granted only after prior assessment of the likely significant environmental effects of these projects has been carried out; whereas this assessment must be conducted on the basis of the appropriate information supplied by the developer, which may be supplemented by the authorities and by the people who may be concerned by the project in question; Whereas the principles of the assessment of environmental effects should be harmonised, in particular with reference to the projects which should be subject to assessment, the main obligations of the developers and the content of the assessment.

Whereas projects belonging to certain types have significant effects on the environment and these projects must as a rule be subject to systematic assessment.

Whereas projects of other types may not have significant effects on the environment in every case and whereas these projects should be assessed where the Member States consider that their characteristics so require; Whereas, for projects which are subject to assessment, a certain minimal amount of information must be supplied, concerning the project and its effects.

Whereas the effects of a project on the environment must be assessed in order to take account of concerns to protect human health, to contribute by means of a better environment to the quality of life, to ensure maintenance of the diversity of species and to maintain the reproductive capacity of the ecosystem as a basic resource for life.

Whereas, however, this Directive should not be applied to projects the details of which are adopted by a specific act of national legislation, since the objectives of this Directive, including that of supplying information, are achieved through the legislative process.

Whereas, furthermore, it may be appropriate in exceptional cases to exempt a specific project from the assessment procedures laid down by this Directive, subject to appropriate information being supplied to the Commission.

2

ECONOMIC VALUE AND SOLUTION OF ENVIRONMENT

INTRODUCTION

Assessing the economic value of the environment is a major topic within the field. Use and indirect use are tangible benefits accruing from natural resources or ecosystem services (see the nature section of ecological economics). Non-use values include existence, option, and bequest values. Those are also called "Intrinsic value" in "Deep Ecology". For example, some people may value the existence of a diverse set of species, regardless of the effect of the loss of a species on ecosystem services. The existence of these species may have an option value, as there may be possibility of using it for some human purpose (certain plants may be researched for drugs). Individuals may value the ability to leave a pristine environment to their children.

Use and indirect use values can often be inferred from revealed behavior, such as the cost of taking recreational trips or using hedonic methods in which values are estimated based on observed prices. Non-use values are usually estimated using stated preference methods such as contingent valuation or choice modelling. Contingent valuation typically takes the form of surveys in which people are asked how much they would pay to observe and recreate in the environment (willingness to pay) or their willingness to accept (WTA) compensation for the destruction of the environmental good. Hedonic pricing examines the effect the environment has on economic decisions through housing prices, traveling expenses, and payments to visit parks.[4]

SOLUTIONS

Solutions advocated to correct such externalities include:

Environmental Regulations

Under this plan the economic impact has to be estimated by the regulator. Usually this is done using cost-benefit analysis. There is a growing realization that regulations (also known as "command and control" instruments) are not so distinct from economic instruments as is commonly asserted by proponents of environmental economics. E.g.1 regulations are enforced by fines, which operate as a form of tax if pollution rises above the threshold prescribed. E.g.2 pollution must be monitored and laws enforced, whether under a pollution tax regime or a regulatory regime. The main difference an environmental economist would argue exists between the two methods, however, is the total cost of the regulation. "Command and control" regulation often applies uniform emissions limits on polluters, even though each firm has different costs for emissions reductions. Some firms, in this system, can abate inexpensively, while others can only abate at high cost. Because of this, the total abatement has some expensive and some inexpensive efforts to abate. Environmental economic regulations find the cheapest emission abatement efforts first, then the more expensive methods second. e.g. as said earlier, trading, in the quota system, means a firm only abates if doing so would cost less than paying someone else to make the same reduction. This leads to a lower cost for the total abatement effort as a whole.

Quotas on Pollution

Often it is advocated that pollution reductions should be achieved by way of tradeable emissions permits, which if freely traded may ensure that reductions in pollution are achieved at least cost. In theory, if such tradeable quotas are allowed, then a firm would reduce its own pollution load only if doing so would cost less than paying someone else to make the same reduction. In practice, tradeable permits approaches have had some success, such as the U.S.'s sulphur dioxide trading program, though interest in its application is spreading to other environmental problems.

- *Taxes and tariffs on pollution/Removal of "dirty subsidies".* Increasing the costs of polluting will discourage polluting, and will provide a "dynamic incentive", that is, the disincentive continues to operate even as pollution levels fall. A pollution tax that reduces pollution to the socially "optimal" level would be set at such a level that pollution occurs only if the benefits to society (for example, in form of greater production) exceeds the costs. Some advocate a major shift from taxation from income and sales taxes to tax on pollution - the so-called "green tax shift".
- *Better defined property rights.* The Coase Theorem states that assigning property rights will lead to an optimal solution, regardless of who receives them, if transaction costs are trivial and the number of parties negotiating is limited. For example, if people living near a factory had a right to clean air and water, or the factory had the right to pollute, then either the factory could pay those affected by the pollution or the people could pay the factory not to pollute. Or, citizens could take action themselves as they would if other property rights were violated. The US River Keepers Law of the 1880s was an early example, giving citizens downstream the right to end pollution

upstream themselves if government itself did not act (an early example of bioregional democracy). Many markets for "pollution rights" have been created in the late twentieth century — see emissions trading. The assertion that defining property rights is a solution is controversial within the field of environmental economics and environmental law and policy more broadly; in Anglo-American and many other legal systems, one has the right to carry out any action unless the law expressly proscribes it. Thus property rights are already assigned (the factory that is polluting has a right to pollute).

ENVIRONMENTAL DECISION MAKING

Public participation practice is a growing part of environmental decision-making. This study looks at its genesis in the latter half of the 20th century. It emerged from an increasing awareness of the fragility of the biosphere in which we all live, coupled with a growing desire to get involved in its protection. Emerging from the Earth Summit in 1992, Principle 10 and Agenda 21 both called for increased public participation in environmental decision-making (EDM) and led to the adoption in Europe of the Aarhus Convention, to which Ireland is a signatory. The legislation resulting from Aarhus is discussed along with its implications. The pertinent legislation is the backdrop, but many other factors affect the development and effectiveness of participatory processes. These factors along with the costs and benefits are outlined. Taking all the relevant factors into account, it is clear that effective participatory processes in EDM require good planning.

The setting of objectives, process planning, selecting the relevant public, and choosing the appropriate methods or tools to use at each stage of a process, are all part of that planning. These are all described together with methods for assessing their effectiveness. In order to see how the methods explored might be applied to a particular process, Environmental Impact Assessment (EIA) is looked at in some detail. International best practice in public participation in EIA is also examined. Methods for assessing the state of public participation on a national level are considered, and a preliminary look is taken at Ireland's status, in the light of all of these. Finally, a survey of Local Authorities, Industry, Environmental Awareness Officers and the Public looks at; attitudes to, experiences of, capacity to deliver, and capacity building measures for, public participation in EDM.

The overall aim of the study was to create a picture of public participation in EDM, as practiced world wide, and to look at what is possible in an Irish context. In order to fulfill the promise of Principle 10 in Ireland, it is necessary to develop effective, creative and worthwhile public participation practice in EDM. This treatise attempts to provide a starting point for the necessary debate required to bring about that promise.

PUBLIC PARTICIPATION IN ENVIRONMENTAL DECISION MAKING

The title of this study, poses several questions: What is the 'public'? What is 'participation'? What are the 'decisions' to which it refers? Starting with the last and working backwards, EDM refers to any process of decision-making where consequent significant environmental impacts are a possibility. This includes law making, planning, strategic planning, resource management planning, licensing of industry e.g. IPPC, environmental assessment (EIA), spatial planning etc. EDM can be even more complex than decision-making on other public issues.

First, environmental impacts do not respect property, jurisdiction or boundaries.

Second, EDM can involve government agencies as both manager and regulator.

Thirdly, environmental issues can provide especially heated value conflicts that require value trade offs.

The level at which the public is involved varies with the relevant legislation, and the attitude of the other stakeholders. Often it just means informing the public of a previously made decision and asking for comments, which may or may not be heeded. Sometimes it means informed consultation. Here there is exchange of information prior to the relevant authority's reasoned decision making, and all inputs are included and seen to be included.

An example of this being the River Basin Management process under Article 14 of the Water Framework Directive as described by Judith Cuff (Cuff, J. 2001). Another example would be the Bantry Bay Coastal Zone Charter. Public participation, at its apex, could also mean that the public itself, in consultation with the relevant bodies, makes the final decision itself. Examples of this can be found under the auspices of Local Agenda 21. For public participation to be effective at any level, it requires the public to be well informed and kept aware of the possibility of participation. This requires a pro-active approach from industry and the relevant public bodies.

What then is 'the public'? The public is often treated as a unitary body, whereas in reality it is a collection of numerous continually shifting interests and alliances (Ortolano), which may be in conflict with each other. The term is used as a "catch-all to describe those with an interest in a decision, other than a proponent, operator, or responsible authority". (Petts and Leach). The individuals making up a public may be involved as individuals or as members of organisations. They may become involved due to their proximity, economics, social or environmental issues, values, etc. There is further discussion on this in Section 5.

By contrast, stakeholders, of which the public is one, are literally those with a stake in an issue and may include non-governmental organizations (NGO's), government or its agents, industry, individuals, communities etc.

Stakeholders do not always want to be involved in an EDM process, but they have the right to know if their interests are affected. They may want to become involved at a different stake of the EDM process.

BENEFITS, COSTS AND THE FORCES AFFECTING PUBLIC PARTICIPATION IN EDM

Public participation in EDM is playing an increasingly important role in the protection of the environment. The forces driving the increased use of participatory practice, the benefits accruing from same, as will as the costs and potential problems are described below.

Forces Driving Stakeholder Processes

Many different forces are at work, prior to, and throughout, a public participation programme. However, the following factors have also moved the participation debate forward. People generally resist change, especially when they do not understand or agree with the goals, the methods, the sponsor or the timing of the proposed change. So, keeping the public in the dark is often a recipe for disaster. Citizens are also increasingly reluctant to defer to 'expert' government agency opinions, and are unwilling to act as sounding boards for bodies that have already made decisions, particularly when they affect their local communities.

Both of these tendencies lead to a lack of public confidence and trust, unless the relevant authorities take note of them. There are growing social expectations for environmental equality resulting in part from the high media profile given to environmental issues. Findings, that minority and low income groups in the USA had been disproportionately affected by pollution and downgraded environments, resulted in the issuing of Executive Order 12898 by President Clinton.

A far-reaching programme ensued, based on meaningful involvement and fair treatment for all races, cultures and incomes in environmental decision-making. "The transmittance of 'experience', rather than knowledge, is the critical neglected dimension of decision-making, particularly as it pertains to our national decision-making"(Barkenbus, J. 1998, p.9). For example development planning in Dublin City requires largely the same expertise as rural development planning, but the experience, vision and values of the rural population may be very different to that of their city cousins. The further the EDM process is from central government, the more pertinent becomes the public participation in same, and in this context, face-to-face meetings are critical to producing real learning and trust.

Previously, with the exception of Development Plans and Waste Management Plans, public involvement in EDM in Ireland was largely restricted to the minimal fulfillment of the two "rules of national and constitutional justice, and in particular 'audi alteram partem' (hear both sides)" (Scannell, Y., Cannon, R., Clarke, M., and Doyle, O. 1999).

In practice this meant that firstly a decision was made and this was then presented to the public for comment, a process also known as 'report and comment'. So, both sides have been heard, but one has the power, and can choose to ignore the input of the other. Changes driven by the Aarhus Declaration and Principal 10 will affect many areas of environmental regulation and planning, opening them up to far greater public participation. Alongside this is the increasing transparency of regulatory processes.

The directive on Freedom of Access to Information on the Environment (Directive 90/313/EC) SI.125 of 1998 and The Freedom of Information Act, 1997 have opened up the EDM processes to closer scrutiny by the public. By compelling the relevant authorities to release documents on request, the inner workings of government have been opened up for closer examination, and relevant information freed up to enable more effective public participation in environmental governance. There is also a growing diffusion of information. The many portals of information available on the internet, together with the traditional media, libraries and the excellent work, in Ireland, of ENFO, all increase awareness of environmental issues.

Since the Aarhus Convention, which agreed to recognise and support the crucial role played by ENGO's in articulating opinions of an environmentally concerned public, European funding has been given to support their activities (Jeffrey, D.W. 2001). □4.62 million was distributed in Europe in 2003. In Ireland, the Department of the Environment and Local Government allocated □132,000 in 2001, and □133,000 in 2002 to build capacity in 23 ENGOs with a national remit.

This was done via a limited company known as Environmental (Ecological) Core Funding Ltd. (Morrisey, P. 2003). It was estimated in 1999 that there were some 500 International ENGOs (Environmental NGOs) world-wide that operated in three or more countries. Apart from the rapid growth of the ENGOs, there is clearly expanded interest in participation. Public unrest at the prospect of development, without formal opportunity for participation in decision-making, is increasingly responsible for expensive delays.

In the UK, policing of antiroad protests has alone cost millions of pounds (Treweek, J.1999). The Shell Oil Company, in the UK, was eventually forced to adopt a major consultation programme on the fate of the Brent Spar, North Sea oil exploration rig, as a result of the international reaction to the campaign waged by Greenpeace. (Acland, A. 2002, 4-1).

Hampshire County Council, in England, with a population of some 2,000,000, had to scrap its waste management plan, following much public unrest, and start from scratch, with a well planned and successful public participation programme. A waste plan was then adopted which had broad support, and is being put into action. (Acland, A. 2002, 4-2). Many industry organizations are realizing that for their good image, public safety, and the longterm financial well-being of their members, responsible partnership with host communities is essential.

There is some evidence that corporations which build sound, responsive relationships with all their stakeholders tend to be longer lived and more successful as businesses in the long term than other corporations (Wheeler and Sillanpaa, 1997). Olson, M. and Toyne, P.(2000) also report a study which suggests a direct correlation between increased public involvement of the public in corporate affairs, and increased profits.

The Chemical Manufacturers Association in the U.S.A. (representing 90% of all chemical production) set up the Responsible Care Initiative, which sets codes of practice. There are six areas covered by these codes viz: Community Awareness and Emergency Response (CAER); The Pollution Code; The Process Safety Code; The Distribution Code; The Employee Health and Safety Code; and The Product Stewardship Code. In relation to all of these, the guiding principles exhort their members "to recognize and respond to community concerns about chemicals and our operations, and to work with others to solve problems created by past handling and disposal of hazardous substances." The adoption, by some industries, of environmental management systems, such as I.S.O. 14001, and E.M.A.S (Eco-Management and Audit Systems) shows their belief that a transparent and responsive industry is also more likely to be a viable one. E.M.A.S. requires public reporting of environmental management and auditing. Internationally, trends towards decentralization in many countries, increases in the numbers of countries with representative democratic forms of government, and conditions laid down by donor agencies on development loans, e.g. The World Bank, are some of the international forces bringing increased public involvement in EDM.

Whilst it is not the intention of the author to go into detail on risk assessment and disaster management., it is worth noting that public participation should, under Council Directive 96/82/EC (Seveso ?) and the Council Decision 1999/847/EC, play an important part in both of these important areas. Wates, N. (2000) gives guidelines for organizing community risk assessment and disaster management programmes, whilst the "UK Guidelines for Risk Assessment and Management" (Defra UK. 2002) give a good introductory insight into the social aspects of risk, and in particular the factors influencing risk perceptions. These factors are listed here and are also relevant to other areas of EDM.

- o Risks which are involuntarily imposed (e.g. pollution from an incinerator) tend to be seen as less acceptable than voluntary ones (e.g. bonfires)
- o Unfamiliar risks (e.g. GMO's) tend to cause concern especially when it appears that they are poorly understood by science.
- o Activities associated with a dreaded form of death (e.g. cancer) are viewed with alarm.
- o Man-made or technological risks (e.g. nuclear power) are less acceptable than natural ones (e.g. Radon). o A risk involving a single large-scale consequence (e.g. civil aviation accident) causes more concern than risks which result in numerous small scale consequences (e.g. car accidents).
- o Alarm may be caused by risks when the consequences of exposure are delayed and cause hidden or irreversible damage (e.g. ionizing radiations).
- o Inequitable distribution of risks and benefits as a result of a particular activity is likely to make the risk less acceptable. An example here might be the remaining operative Chernobyl nuclear power stations. Employees, knowing the risks involved but needing to earn a living, would have a different view

of risk to others living in the vicinity, and different again to those living 100 kms away, and relying on the electricity to heat their homes in temperatures of-30°C.

- o Activities posing a threat to certain groups such as children or future generations are generally more worrying.
- o Risks which are the subject of controversy and contradictory information generally cause concern.

Benefits of Public Participation.

The protection and enhancement of the environment is the main aim of most EDM processes, and public participation can improve the quality of both the process and the end decision. The participation process is also a learning one for all the stakeholders involved, especially if there is a free flow of information between the parties.

For the public it can often be the first experience of taking an active part in the democratic process. EDM processes benefit from the direct and immediate knowledge held by citizens and business, concerning environmental conditions in their communities and industries. Encouraging the public and other stakeholders to share their knowledge, with the regulatory authorities, fosters better-informed decisions and decreases the likelihood of environmental harm, whilst increasing project viability.

Of 25 overseas projects sponsored and evaluated by the World Bank, 13 failed mainly through lack of local input (E.L.I. Research Report, 1997) (Technical Document 1, 1996). It should be realized that "wisdom is not limited to scientific specialists and government officials," (120 p.6) and that "rational analysis, carried on in ignorance of political reality, may well end up so divorced form social reality, as to be of little use to anyone." (Barkenbus, J. 1998, p.2) (Acland, A. 2002).

Public participation gives broader perspectives on a particular process, and early involvement gives: added time to study issues and develop the process; enhanced credibility of the decision making process; early identification of the diverse perspectives on the issues of concern; and the generation of solution options (E.L.I. Research Report, 1997). Better-designed projects, which avoid costly delays in appraisal and implementation, can also result from early and planned consultations and public participation (Bisset, R., 2000) (Acland, A. 2002, 4-1 to 4-5).

For the proponent, early participation can have the added benefit of diffusing opposition to a project. If a broad based consensus is built, it can also lead to a public sense of 'ownership' (Acland, A. 2002, 4-2). The public's enthusiasm is potentially a powerful motivating force for the EDM process. Public input can also supplement scarce government monitoring, inspection, and enforcement, resources (Bisset, R. 2000).

Involvement in EDM is a learning experience giving the public insight into the governance process. If the experience is good, it powers the way for future co-operation. The converse can also be true! Experience has also shown that environmental issues are powerful catalysts of civic 9 participation and serve as a good incentive for citizens' action and responsible democracy (Technical Document 1, 1996). Executive Order 12898 introduced by President Clinton is a far-reaching programme, based on meaningful involvement and fair treatment for all races, cultures and incomes in environmental decision-making.

It has done much to prevent the dumping of dirty industry on the marginalized members of the USA citizenry. The order requires state agencies, and in particular the EPA, to pro-actively seek out the disenfranchised members of communities likely to be affected by developments, or already living in depleted environments, and encourage them to become involved in the relevant EDM processes.

Costs of Public Participation

There is limited information on the relative costs of using the different methods of participation practice, although some attempt is made to quantify these in the section 5.4. which describes each method. Actual costings for specific processes are given in a number of studies (Petts, J. and Leach, B. 2000) (Cuff, J. 2001. pps 85, 89 and 97). Clearly deliberative processes, which are used to engage relevant stakeholders in debate, discussion and deliberation, could only be done as part of a major project. Expenses range from □20-40,000 (for a citizen jury with 16 participants for 5 days), to □200-300,000 (for 3 community advisory groups of 16 people over 6 months). (Petts, J. and Leach, B. 2000, P. 41).

However, extensive use of participatory processes in the 'Wise Use of Flood Plains' project showed costs of about 10% of the above. Most of this funding was used to employ facilitators (Cuff, J. 2001). The costs to members of the public in time and money, as well as emotional stress in situations where access to EDM is limited, do not appear to have been estimated previously. An attempt is made to gather some facts about the first two in this study. The intensity of the stakeholder process, including commitments of time, energy and money, and the often, uncertain results, has led to a burnout phenomenon amongst many participants, and from all stakeholder groups. (Yosie, T.F. and Herbst, T.D. 1998) Excessive raised expectations about participation can lead to frustration and anger with a breakdown in trust. It is essential then to be clear with all participants as to the limits of influence of the process.

Where an infrastructure project crosses planning jurisdiction boundaries or where more than one regulator is involved, the costs to the public, state and proponents are multiplied. The increased time scale and expense involved in public participation is of particular significance in projects with a small capital funding. An international example of this would be the AIJ projects (Activities Implemented Jointly under the Climate Change Convention). These projects already require appraisal by two governments and then there is the extra of public participations costs. After this they must compete with traditional energy supply systems, where often no regulatory costs are involved.

Problems Encountered in Public Participation Practice

The main areas of difficulty affecting public participation are attitudinal, lack of capacity to deliver programmes, lack of clarity about what outcomes are possible and the lack of a legislative framework. Some aspects of these issues are addressed below. The difference between success and failure is not always clear. If the public participation process delivers a clear decision, which is then overruled by the legislature, is this a success? (E.L.I. Research Report 1997 p. 13). A good example here is the Bantry Bay Integrated Coastal Zone Management (ICZM) Project, funded for three years under the EU Life Programme. After three years of intense community activity and the drawing up and enactment of a text-book public participation plan, the Bantry Bay Coastal Zone Charter was created.

At this point, no further funding was available, and the Project was closed down. Clearly the process was a success, but the result not. There is a general lack of experience of participatory processes in Ireland, and very few stakeholders have experienced planned participation processes. There may be suspicion, cynicism, or enthusiasm, but there is unlikely to be previous training. Suspicion about participation could be from politicians, who may feel their power is being diluted, or from ENGOs that have very specific and fixed viewpoints. Cynicism could be from previous public participation processes which were poorly managed, leaving participants reluctant to try again (E.L.I. Research Report, 1997). Citizens are frustrated when they are treated as adversaries, rather than welcome participants in the EDM process........ they feel they have been invited too late in the process. Public 'hearings' often do not include 'listening'. Citizens efforts and ideas are not included in proposals and no reason given.

The resources available to proponents and government are overwhelming. Litigation and direct action at least get a reaction" (C.E.Q. 1997. p.18). Similarly, public hearings are criticised for: being held at times inconvenient for the public, establishing an atmosphere that inhibits dialogue, and conducting proceedings that intimidate the public (Lein, K.L. 2003, p.56).

The lack of a clearly defined purpose can also means that the role of the public is sometimes unclear not only to the public but to those delivering the process, with the consequence that the participation process chosen is often unsuitable for the particular EDM process (Breggin, L and Hallman, H. 1999. p.44). The making available of the necessary resources for a worthwhile public participation process, whether by the proponent or the regulator, by the commitment of personnel and finances, is essential. The implications for project proponents, of not having a properly planned and funded participatory EDM process, may prove much more costly, than dealing with the backlash. In many circumstances, the choice increasingly is not whether to involve the public, but how to get the best value from the chosen process (Petts, J. and Leach, B. 2000).

To avoid hearing from only the activist or the powerful elite, and in order to get the widest sweep of opinion and information, authorities must reach out into the community (Canter, L.W. 1996. p.592). In this regard some activities are best avoided, for example public meetings.

In the absence of trained facilitators these can often come down to 'he who shouts loudest wins!' Lack of technical support for the public, and difficulties in getting access to information can diminish the ability of the public to play a meaningful part in EDM processes. Public access to independent expertise in a particular technical/scientific field can prove difficult, particularly where the proponent of a project is a monopoly employer in the industry in the state. Faced with five weeks to object to a planning proposal, for example, this is a major hurdle. Added to this, it is the experience of the author that stakeholders with no technical/scientific background find it problematic to access, comprehend or evaluate data and information, especially under these short time constraints.

The feelings of powerlessness engendered by these hurdles, add to public perceptions of a lack of influence at both lower and higher levels of national and European government. "Consultations of interested parties.... can only ever supplement, and never replace, procedures and decisions of legislative bodies which possess democratic legitimacy" (COM (2002) 704. final p.4).

This sense of powerlessness may be part of the reason why, even in cases where considerable energy and resources are expended, to identify individuals and groups, only a small proportion of the public ever attend participation programmes (Petts, J. and Leach, B. 2000. p.18). Public involvement may actually result in an increased level of conflict (134.p18). Becoming involved in a consultative process where the decision has already been made, or where the possible outcomes are not made clear at the outset, can lead to a great deal of frustration.

Anger may follow when the input of the public is ignored, especially following prolonged constructive engagement with an EDM process. Conflict can also occur between professionals. One source of such disagreements could evolve from the different emphases of 'scientific peer review' and 'social peer review'. The scientific peer review process is well established, and is essential to assess the technical information provided to policy makers.

The parallel social peer review, designed to obtain societal acceptance and legitimacy for the decisions rendered, by contrast, does not have a set of recognised professional standards. (E.L.I. Research Report. 1997. p.14). It should also be remembered that whilst science may strive for the 'truth', the opinions of scientists are coloured by their values and beliefs. Scientists, engineers and other 'experts' should show respect in the face of 'emotional', cultural and traditional forms of wisdom. The perceptions created by a technocratic philosophy encourage NIMBY activism and further polarise communities.

The resulting deadlock frustrates the ability of planners to achieve consensus on initial issues and contributes to anti-participation attitudes. Therefore, rather than define 'NIMBY'attitudes as an irrational response to problems ordinary citizens cannot grasp, the solution is to encourage more participation not less (Lein, K.L. 2003. p.56). The misconception that the public are overly emotional, ill-informed and solely motivated by selfish interests, in their obstruction of facilities for the common good, leads to an information-based strategy.

The assumption being that, 'if only the public knew the truth!' This assumes the existence of an objective truth to know (Petts, J. and Leach, B. 2000. p.9). The conflict that arises from having different bodies making decisions on different parts of the same EDM process can often confuse the public understanding and increase frustrations. For example the executive split of planning and environmental matters in some aspects of development between the Local Authority and the EPA. Public involvement in these circumstances must deal with two different agencies (Canter, L.W. 1996. p.592) (Gray, A.J. et al 2002. p.6).

Organisations resist change, and communication blocks within organisations can prevent them from making timely responses to their publics. The established routines and organizational systems of many bodies seek to promote the 'status quo'. Organisational structures which support sustainability, and therefore public participation, are a threat to 'command and control' style management systems. Managers, who have risen to power in this style of structure, frequently resist the transition to alternative structures, which embrace public consultation (Griffiths, A. 2000). Many opportunities for public participation are laid down in the environmental legal framework, as shown in Appendix 1. Where legislation does not specify early public involvement, then it is at the discretion of the relevant authority, and dependent on the opinion of the relevant public servant.

For example, in the development planning process, the Manager's reporting of the consultative process clearly includes the names and nature of all submissions together with his response to them (Section 11 of the Planning and Development Act 2000), whereas in making an Air Quality Management Plan (under Section 48 of the Air Pollution Act, 1987 and Section 102 of the Environmental Protection Agency Act 1982) there is no legal compulsion to show how any submission might have affected the final decision.The recent use by Cork City Council of Article 81 of The Planning and Development Act, Regulations, 2001, to enter into a public/ private partnership with a company previously refused planning permission to develop the same site (TP 02/26234), leaves the public only minimal scope for comment and no route to appeal any decision made.Under section 19 of the Protection Of The Environment Act, 2003, the making, variation or replacement of a Waste Management Plan, has been made an Executive Function. This has removed the immediate political accountability of the decision-making, and so removed another route for public input

THE HISTORICAL PERSPECTIVE

General provisions for public access to information, participation and justice go back 14 centuries in Islamic

Law, as recognized in many countries of North Africa, The Middle East and Asia. (Ahmad, A., 2002) but the International Boundary Waters Treaty, 1909, was the first global treaty to include strong provisions for public access to information and participation (Bruch, c. and Filbey, M., 2002). Broad international recognition of public involvement including provisions for transparency, participation and justice dates back 55 years. The Universal Declaration of Human Rights in 1948 provided the kernels for generalized access to information (Article 19) and Justice (Articles 8 and 10) as well as the right to associate (Article 20). In the West widespread public awareness of changes in the balance between human life and the environment emerged in the1960's, with the publication of Silent Spring (Carson, R. 1962), The Population Bomb (Ehrlick, P.R.,1968) and Small is Beautiful (Schumacher, E.F., 1973).

The immediacy of television portrayals of such disasters as the Torrey Canyon and "acid rain" (Jeffrey, D.W., 2001) made an even greater impact. In the USA as in much of the West, the 1960's and 1970's were marked by a participation movement (Breggin, L. and Hallman, H., 1999) which led to legislative changes. The Freedom of Information Act, 1966 gave access to data and information held by government agencies. The National Environmental Policy Act, (NEPA) 1969 amongst its measures required public review of Environmental Impact Statements (EIS). Previously the Administrative Procedures Act of 1946, directed government agencies to introduce a notice and comment stage into rule making procedures and to keep registers.

This Act also allowed for the right of appeal to federal courts (Ortolano, L., 1997) So together these three pieces of legislation provided the foundation for the involvement of the public in environmental protection in the USA, and NEPA has largely been the basis on which environmental impact assessment world-wide has been based. The voice of environmental concern was internationalized at the

Stockholm Conference, 1972 (United National Conference on the Human Environment), drawing together both developed and developing nations, but in the absence of the Soviet Unions and most of its allies. For the first time environmental non-government organizations (ENGO's) played a important part in the deliberations.

Stockholm was the event that turned the environment into a major issue at the international level, declaring the right of people to live "in an environment of a quality that permits a life of dignity and well being". (GEO-3. UNEP., 2002) Amongst its impacts were:

- Some 50 governments together with the Organisation for African Unity (OAU) have recognized the environment as a fundamental right.
- During 1971-1975, 31 major national environment laws were passed in OECD countries, compared with 32 in the previous 15 years.
- By 1982 the number of countries with environment ministries rose from 10 in 1972 to 100.

Although the link between environment and development was made in Stockholm, it was the 1983 World Commission on Environment and Development (WCED) also known as the Bruntland Commission that first developed the idea of Sustainable Development, in its broadest sense. The WCED report in 1987 resulted from consultations at all levels of the global society, and attempted to balance the needs of future generations with those of our own. In essence it tried to promote both the inter and intra generational equity (GeO-3 UNEP., 2002).

However, in its concluding statement, the commission stated that "existing decision making structures could not cope with the demands of sustainable development" (WCED.1987). The UN Conference on Environment and Development (UNCED or "The Earth Summit") held in Rio

De Janeiro, Brazil, saw 176 government, 100 heads of state, 10,000 delegates, 1400 NGO's and 9,000 journalists gather in what is still the largest event of its kind ever (GEO-3.UNEP., 2002). The summit was preceded by global consultations with hundreds of thousands of groups and individuals.

Amongst its many achievements, the two most relevant here are, Agenda 21, and Principle 10 (P.10), of the 'Rio Declaration on Environment and Development'. Agenda 21, in its 40 chapters, lays a solid foundation for the promotion of sustainable development. Chapter 8, Section 1, calls on governments to seek broader public participation in policy making and decision making for sustainable development. Section 3 is devoted to the strengthening of the roles played by the principal social groups: women, children and young people, indigenous populations; NGOs; local government; workers associations and unions, business and industry; and the scientific and technological community (NCEDR, 2003) (Agenda 21, Chapter 8.).

Chapter 40 establishes two programme areas to ensure that decision making is based on sound information: namely, bridging the gap in availability, quality, coherence, standardization and accessibility that exists between developed and developing countries; and improving availability of information. Principle 10 states:

"Environmental issues are best handled with the participation of all concerned citizens, at the relevant level. At the national level, each individual shall have appropriate access to information concerning the environment that is held by public authorities, including information on hazardous materials and activities in their communities, and the opportunity to participate in decision-making processes. States shall facilitate and encourage public awareness and participation by making information widely available. Effective access to judicial and administrative proceedings, including redress and remedy, shall be provided".

It would be difficult to overstate the significance of Principal 10 in providing a mandate to regional and international governance development in the decade following the Rio Declaration, as well as to an underlying commitment motivating national law making. Since 1992 a raft of regional and international treaties have incorporated P.10, whilst addressing such issues as Desertification, Bio-Diversity and Clean Development Mechanisms. International organizations such as the World Bank, and to a much lesser degree the World Trade Organisation, have made steps towards making public consultation and involvement an integral constituent of their development loans.

In the context of this study and from an Irish and European perspective, the most important initiative evolving from P.10 is the 1998 UN/ECE Convention on Access to Information, Public Participation in Decision-Making and Access to Justice in Environmental matters. The ECE (Economic Commission for Europe) is a regional body of the United Nations and includes 55 countries. The European Community plus 39 other nations have signed the convention (also known as the Aarhus Convention), which it was ratified in 2001. The Aarhus Convention is based on the three pillars, ensconced in P.10 and elaborated in Agenda 21. The pillars are; access to information; public participation in decision-making; and access to justice.

The Aarhus Convention

The Aarhus Convention sets out to put a shape on the aspirations of Principle 10 (P.10). It encapsulates the three main pillars of P.10, access to information, public participation in decision-making, and access to justice, and sets the basic tenets for all European legislation on EDM. Pillar I-Access to Information. Article 4 of Aarhus sets out the general right of persons to gain access to existing information, on request. Pillar II-Public Participation in Decision-Making. Article 6 covers public participation (PP) in decisions on specific activities, e.g. on the proposed locating, construction and operation of large facilities. Article 7 deals with PP in the development of plans programmes and policies relating to the environment, which includes sectional or land use plans, environmental action plans, and environmental policies at all levels. Article 8, covers public involvement in the preparation by public authorities of laws and regulations.

In each of the Articles, 6, 7 and 8, early involvement of the public is encouraged. Articles 7 and 8 are less precise than Article 6. Article 6 provides for a high level of involvement adequately guaranteed by law (Stec, S. and Casey-Lefkowitz, S., 2000. p.86). Article 6 can apply to spatial planning decisions, development consents, operating permits, discharge permits, or any particular proposed activity where the decision making may have a potentially significant impact on the environment. Article 6 is not limited then, to activities where an EIA is required or to those activities listed in Annex 1 of the Convention.

Pillar III-Access to Justice

Article 9 allows the public to challenge "acts and omissions" by either public authorities or private persons, as applied to articles 4 and 6, as well as allowing them a review of violations of domestic environmental laws. In relation to Article 6, members of the public must have a significant interest {Article 9 (Section 2(a)} or maintain impairment of a right (Article 9 section 2 (b)). A level of standing is given to both individuals and NGO's. The provisions of an expeditious, free or very cheap, alternative review mechanism established by the courts is to be made available.

Publicity about these procedures and removal of barriers to access are also stressed. It is clear that in order for Pillar II to operate, Pillars I and III must also be well established, i.e. access to decision making cannot be effective without good information resources and the possibility of enforcement through access to justice. It should be noted that Article 3 reminds the parties that the provisions of the convention are only minimum requirements.

The Implementation of Aarhus

The implications for environmental governance under European environmental law and consequently that of the Republic of Ireland are far reaching as will be seen below. Pillar I-Access to Information. National Level. The Freedom of Information Act, 1997, (FOIA) gives access to all information held by the state, including information on the environment, other than the exceptions listed in the Act. ENGOs, however have told the author that they see S.I.125 of 1998 as a more effective tool. Directive 90/313/EC (on public access to information on the environment), implemented in Ireland as S.I. 125, of 1998, (The Access To Information On The Environment, Regulations, 1998.) has already been repealed by the European Commission and replaced by Directive 2003/4/EC, which must be brought into Irish law by 14th February 2005. The resultant changes include:

- Persons or bodies which perform functions by arrangement with a public authority and which effect the environment are themselves deemed to be public authorities for the purposes of the Directive (Article2, Section 2(c)), and so should make available environmental information held by or for them on request.
- Applicants for information do not have to state an interest.
- Information should be made available in the format requested (Article 3, Section 4).
- "Course of Justice" exceptions refer to information that may adversely affect the course of justice (Article 4, section2 (c)) rather than the previous "sub-judice" exceptions.

- Definite response times and appeals procedures are laid down. EC Institute Level.

Regulation (EC) 1049/2001 grants European citizens and residents a right of access to documents of the European Parliament, Council and Commissions. Applicants receive an acknowledgement of receipt, and within 15 days provide the item(s) requested or give reasons why not. The applicant is entitled to complain to the Ombudsman or institute Court proceedings if a negative reply is received. 20 pages of A4 text will be photocopied free of charge and any extra charged at cost.

Pillar II. Access to Participation

National Level

The proposal for a Directive of the European Parliament and Council providing for Public Participation in respect of drawing up of Certain Plans and Programmes relating to thc Environment, and Amending Directives 85/337/EC as implemented in Ireland by S.I. 349of 1989 (and amended by Directive 97/11/EC), and 96/61/EC.

These two refer to "Environmental Impact Assessment" and "Integrated Pollution Prevention and Control", respectively. The new directive will make changes in PP procedures required by these directives as well as creating a requirement to consult the public in drawing up of plans or programmes in seven other directives viz:

- The 1975 Framework Waste Directive (75/442)
- The 1991 Batteries Directive (91.157)
- The 1991 Agricultural Nitrate Directive (91/676)
- The 1991 Hazardous Waste Directive
- The 1994 Packaging Directive (94/62)
- The 1996 Ambient Air Quality Directive (96/62)
- The 1999 Waste Landfill Directive (99/31)

Honouring the subsidiarity principle, the proposed directive sets out the minimum requirement for effective public participation, in fulfillment of the objectives laid down by the EC Treaty in the field of the environment, leaving the detailed arrangements to the member states. The proposed directive requires member states to take the necessary measures, to ensure that the public are given early and effective opportunities to participate in the preparation (and review) of the plans and programmes listed above. For this purpose, "the public" has the same meaning as in Article 2(4) of the Aarhus Convention. To that end, Member States are required to ensure that:

- o the public are informed (by public notices or other appropriate means) about any proposals for such for such plans or programmes and that relevant information about such proposals is made available to the public.
- o the public are entitled to express comments and opinions before decisions on the plans and programmes are made;
- o and that in making those decisions, due account is required to be taken of the results of the public participation.

Taking the EIA directive 85/337/EC as an example, the following changes are proposed. The Aarhus Convention definitions of "the public", and of the "public concerned" are inserted into 85/337/EC. The public concerned will be defined as the public affected by or likely to be affected by, or having an interest in, the development consent procedure. A new Article 6(2) requires Member States to take steps to ensure early and effective opportunities to participate. The new Article 6(3) will lay down requirements for informing the public based on Article 6(2) of the convention. This requires the disclosure of all pertinent documents and opinions expressed by relevant public authorities. The new Article 6(4) requires Member States to provide proper opportunities to express comment and opinions as early as possible, but definitely before a decision is taken.

Reasonable time frames for the stages of participation should also be provided. Changes to Article 7 of the EIA Directive regarding Transboundary EIA are discussed. The IPPC Directive will have amendments similar to those of the EIA Directive, with the addition of a new Annex V in which detailed provisions for public participation will be set out. Both the Water Framework Directive (2000/60/EC) and the proposed Directive 2001/42 on "the assessment of the affects of certain plans and programmes on the environment (SEA Proposal), already include provisions which are consistent with Article 7 of the Aarhus Convention (Banfield, N., 2003.). EC Institute Level Although no legislative proposal exists at present, COM 2002/704 Final, entitled, "Towards a reinforced culture of consultation and dialogue-General principles and minimum standards for consultation of interested parties by the Commission", provides good insight into the governance practice being developed within EC structures.

By fulfilling its duty to consult (Protocol No.7 annexed to the Amsterdam Treaty), "the Commission will help to improve the quality of policy outcome and at the same time, enhance the involvement of interested parties and the public at large". (COM.2002.p.5). The commission has underlined its intention to "reduce the risk of the policy makers just listening to one side of the argument or of particular groups getting privileged access" (COM.2002 p.5). These minimum standards laid down in the communication were themselves the result of a European wide consultative process.

Pillar III Access to Justice

National Level

Both Directive 2003/4/EC and the proposed Directive 2000/0331(COD) amending 85/337/EC and 96/61/EC, contain provisions for access to justice in line with the Aarhus

Convention. Changes to the other directives listed are given in Annex 1 of the proposed directive 2000/0331(COD). However, shortcomings in the enforcement of environmental law throughout Europe, as outlined in the Sixth Environment Action Programme (6EAP), (Com (2002) 31, Final), will be the subject of a new Directive being worked on by the commission at present (Banfield, N.,2002).. All of these measures will become part of Irish Law in due course, when transposed by the Dail. In 6EAP it is recognized that better access to courts for NGO's and individuals would have a number of helpful effects in relation the implementation of community law. EC Institution Level.

Rules on access to justice with regard to community institutions are laid down in the Treaty of Rome. The Aarhus Convention will have to be ratified by a legislative instrument, taking account of existing provisions, in order to ensure compliance at Community Institution level.

In Johannesburg, South Africa 2002, a new initiative called Partnership for Principle 10 (PP.10) was launched by governments, donor groups and NGO's, as one of the outcomes of the World Summit on Sustainable Development (WSSD) (WRI Press release, 2002). In UN parlance this is a Type 2 outcome, meaning it involves non-governmental stakeholders, whilst being linked to inter governmental commitments, in this instance Principle 10, (P.10) and WSSD Plan of Action.

In PP.10 each partner makes specific commitments based on the promotion of P.10 at national and international level, by identifying implementation gaps, by capacity building, and by improving and implementing relevant legislation. Each partner is held accountable for fulfilling its commitments. PP.10 strategy is to support independent assessment by NGO's of their countries performances based on a common framework. Independent assessment by NGO's using systems such as The Access Initiative (Petkora, E., Maurer, C., Henninger, N. and Irwin, F.,2002.) are to be received by government agencies and other stakeholders and priorities for action identified. The European Commission has committed itself to PP10.

Transboundary Environmental Impacts

The Convention on Environmental Impact Assessment in a Transboundary Context (the ESP00 Convention), was signed by the European Community in 1991. This initiated changes incorporated into the EIA Directive 97/11/EC and the IPPC Directive 96/61/EC, creating provision for cross border public consultation. Article 7 of the EIA Directive requires member states to send information 'as soon as possible' and not later than when its own public is informed. The affected state must be sent a designation of the project together with any available information on its possible transboundary impact, as well as information concerning the nature of the decision to be taken.

If the affected state wishes to participate in the assessment process, it must also be sent the application submitted by the developer under Article 5, together with the information relevant to the assessment procedure. All the above information must be made available to the public in the affected State, should the latter decide to participate in the assessment process.

Within a reasonable time scale, submissions must then be accepted by the competent authority in the Member State, in which the development is proposed. The notifying Member State under Article 8 must take the transboundary consultations into account in making a decision, and under Article 9 must inform the Member States consulted under Article 7 of the decision made. With regard to the decision there is no compulsion to notify the public in the affected member states, even if they make submissions to the transboundary EIA process. (Macrory, R, and Turner, S., 2001). Article 17 of the IPPC Directive, drafted in parallel with the 97/11 Directive is less explicit with regard to the provision concerning cross border public consultations and it is proposed to amend both to bring them in line with the Aarhus Convention.

With regard to transboundary EIA, these proposed amendments will ensure equal rights of consultation for the public whether they live in the notifying or affected Member States. However, in the case of the IPPC permitting process, the public of the affected state will remain entitled only to access the IPPC application itself (Macrory, R. and Turner, S. p.14). The Water Framework Directive is still the only measure that includes not just transboundary consultations but also power sharing (Article 14(i).) Regardless of the above, it would appear that Ireland had failed to implement fully the 1985 and 1997 EIA Directives with regard to Transboundary consultations, as ruled by the European Court (Case c-392/96 (1999) CMLR 727). The Planning and Development Act 2000 whilst it empowers the Minister for the Environment to make regulations regarding transboundary EIA, it 26 does not give rights of consultations for the Irish public nor that of other Member States (Macrory, R. and Turner, S. p.17). As can be seen from the above, there is a great lack of clarity on the issue of consultation with regard to transboundary environmental issues.

CAPACITY BUILDING

Defined by Wates, N. (2000) as "the development of awareness, knowledge, skills and operational capability in order to achieve a purpose", capacity building is an essential part of the development of public participation practice, and it is necessary to build capacity amongst all the stakeholder groups in society.

Agenda 21, Chapter 36 set out the broad proposals for promoting education, public awareness and training which it sees as meeting capacity building needs. It promotes the inclusion of environmental and human ecological considerations in all aspects of education and decisionmaking.

'The Access Initiative' (Coyle, J. et al. 2002) in a study of nine nations found that: lack of government capacity constrains public access e.g. through lack of staff, equipment, procedures and training. Laws and procedures relating to access are often new, and the necessary training has not been done to facilitate their enactment. South Africa was seen as the only country, of the nine studied, with an extensive programme building staff capacity. Lack of capacity in civil society was also seen as a serious impediment in all the countries studied.

The Environmental Law Institute (Breggin, L. and Hallman, H. 1999) looked in detail at Building Capacity to participate in US EPA activities and pinpointed the following as essential elements in capacity building.

- o Information.
- o Technical Assistance.
- o Process Education.
- o Access to Documents.
- o Education on laws as fundamental building blocks.

The study then went on to give potential approaches to enabling stakeholders to participate, showing models of each method in action. These include;

- Independent Information Broker
- Ombudsperson
- Hotlines
- Technical Assistance Grants
- Citizen Training on Processes and Legal Requirements
- New Collobrative Participation Processes
- Increased Date Availability and Dissemination Networks
- Grants to Community Groups
- Improved Access to Documents
- Improved Mailing Lists

Capacity building in the Irish context will be looked at in the following chapters, but the following are a few examples of best practice at work elsewhere. The US EPA's 'Community Based Environmental Protection' programme is another way to create local ownership of the environment (US. E.P.A., 1999).

The many Local Agenda 21 projects worldwide are achieving similar results. In the UK, through the Environment Council, training programmes for all stakeholder groups are available on stakeholder dialogue, as a tool to achieve sustainability. Information tools to enable local government employees to respond to their publics now include a one-stop internet portal.

In Ireland, IBEC runs regular training programmes for people in Industry, on 'Environmental Communication', to promote better communication with the other stakeholders, regulatory bodies, the media, and the local community (IBEC, 2003).

IRELAND AND PRINCIPLE

The Access Initiative

Over the last three years the World Resources Institute has formulated a methodology to assess the implementation of Principle 10 at national level, called The Access Initiative (TAI) (The Access Initiative, 2003). This is a global coalition of civil society groups collaborating to assess and promote national implementation of Principle 10 commitments. The European Commission has now signed up to Partnership for Principle 10 (P.P.10), which commits governments to support such independent assessments, by NGO's, of national Principle 10 performance.

The pilot project, reported in 'Closing the Gap' (Coyle, J. et al. 2002) involved nine nations on five continents and twenty-five civil society groups. Following this, the process was refined, and the difficult task begun of finding a methodology for assessing the implementation of Principe 10 in the many different cultural and political systems that exist. The methodology looks at:

(i) The comprehensiveness and quality of the General Legal Framework

(ii) The degree of available access to selected types of information about the environment.

(iii) The degree of public participation in decision making processes, and

(iv) Comprehensiveness and quality of capacity building efforts to encourage informal and meaningful public participation.

In order to conduct an assessment of the Irish government's performance, and so to pinpoint weaknesses, with a view to capacity building, the Access Initiative requires the building of a coalition of civil society groups within Ireland.

The formation of a coalition is necessary to get the skills and knowledge base necessary to perform the research and analysis tasks contained in a full Access Initiative Assessment. Potential coalition members might include NGO's, lawyers, academics, journalists, etc. The structure of the coalition would include a local organization, a research team with membership from across the coalition, and an advisory group of 'experts' to oversee the research and give credibility to its findings (The Access Initiative, 2003).

The work involved, with 6-12 researchers, is estimated to take 4-6 months, based on the pilot project. Clearly the degree of detailed research required to carry out a complete TAI study of Ireland is beyond the scope of one individual and of this study, but a selective look at the issues raised will give some insight into the current state of environmental governance in the Republic.

Access to Information

Law: Whilst there is not constitutional right to information on the environment, freedom of access to information is available under the Freedom of Information Act, 1997 (FOIA) and the European Directive 90/313/EC (S.I. 125 of1998) (replaced by 2003/4/EC). Access under FOIA is restricted by cost, limiting its use by people with low incomes. Emergencies: Although a general legal mandate exists (The Seveso Directive) for dissemination of information about environmental and health impacts to the public during and after an emergency, An Taisce has repeatedly expressed its concern about the implementation of same (Lumley, I. 2003)

Monitoring: In general, monitoring of air and water quality is carried out to a high standard and information is made available about the results of same. Both air and water quality monitoring results are available on the EPA website, and are freely available to the public through ENFO. The quality of this information is of a good standard, but the data is 1-2 years old. Local Authorities also carry information on their web sites, though the range and quality varies considerably.

State of the Environment Report (SOE): The EPA, as mandated by the EPA Act 1992, produced its third SOE Report in 2000, and it is rich in data and indicators to describe conditions and trends. The report was well publicized in the media, and available in bookshops. 'Government in Focus-Key Environmental Indicators For Ireland 2002', is available on the web-site.

Facility-level Information: Readily available and timely information on the environmental impact of particular facilities is essential to enable people to protect their environment. In the USA online information regularly updated is available on the Toxic Releases Inventory (TRI) web-site. In Ireland the EPA is developing a similar system, based on GIS mapping. Local Authorities have some information on web sites, but not facility specific. Otherwise it is necessary to trawl through registers at EPA or Local Authority offices.

Participation in Environmental Decision Making

Law: The legislative and constitutional frameworks regarding participation are beyond the scope of this study, but would be addressed in an Access Initiative Assessment. Whilst public participation is allowed at an early stage in some environmental decision making processes (e.g. Development Plans) in others it happens too late to make substantive changes (e.g. the EIA process), and in others, there is no right of appeal to an independent arbiter such as An Bord Pleanala (e.g. Article 81,Planning and Development Regulations, 2001). Policy: Lead times for notification of public participation are acceptable in terms of the Access Initiative, though the author's experience would suggest that this depends on the effectiveness of the notification procedure. There is a wide availability through ENFO, the Internet, Local Government Offices, Libraries, etc. of policies, strategies, plans programme and laws.

The requirement to send development plans, planning applications etc. to the 'statutory bodies' gives a degree of external consultation by non-involved expertise, but as explained elsewhere, much environmental decision making in Ireland is of the 'report and comment' type. Project and Plans: The lead in time for public comment on draft plans is generally eight weeks, but it is five weeks for a planning application, regardless of the project size and complexity. The quality of information provided to the public and relevant authorities, particularly EIS's is very varied, and is often, in the author's experience, poor both in content and presentation. Clearly this does not allow for good decision-making. Records of local permits and other project documents are accessible at the relevant offices and increasingly on the internet.

Capacity

Law: The right of freedom of association and the right to a clean environment are guaranteed under the Irish constitution.

Government: The establishment of FOI officers in all the government departments dealing with environmental issues, and the publication of the EPA Section 15 and Section 16 reference books are some of the ways in which capacity is being built by the government with regard to access.

The presence of ENFO (The Environmental Information Office) as a one-stop shop and library for all environmental information is a very important capacity building force. ENFO also provides support for schools and materials for environmental education.

ENGO's are encouraged and give some funding under European Parliament Decision No. 466/2002/EC. It is not intended to give further comment here on access to EDM in Ireland as this will be discussed later in the light of the results of the survey undertaken as part of this study.

ENVIRONMENTAL IMPACT ASSESSMENT (EIA) AND PUBLIC INVOLVEMENT

EIA is a gateway control and not in itself a permitting or authorisation process. It is a very important tool to enable good decision-making, and although its findings usually correlate to the final decision, theoretically the reverse can be true. Article 6 para 2(e) of the Aarhus Convention states that "the fact that the activity is subject to a national or trans boundary impact assessment procedure", dictates that the public concerned shall be informed, either by public notice or individually as appropriate, early in an adequate, timely and effective manner".

Further, Articles 4 and 5 require each Party to provide for public participation when all options are open, whilst encouraging applicants to identify the public concerned and enter into discussions prior to applying for a permit.

Within Ireland Council Directive 85/337/EEC as amended by Directive 97/11/EC lays down the requirements for the EIA process, and is given effect in Irish Law through the Planning and Development Act 2000 (implemented by The Planning And Development Act, Regulations, 2001.), through IPPC Licensing and Waste Licensing and through the European Communities (Environmental Impact Assessment) Regulations, 1989-1999. (EIA Regulations) (Doyle, A., 2002).

In the EIA Regulations, a developer when required must submit an Environmental Impact Statement to the competent Authority. An EIS is defined in S.I No. 349 of 89 (Article 3 (1) as follows " A statement of the effects, if any, which a proposed development, if carried out, would have on the environment". Following the submission of an EIS, the public, relevant public bodies and, where legislation requires, other governments, have the right to comment.

The competent authority must take all submissions into consideration and make a decision. The decision and the reasons for taking it must then be communicated to the public and other interested parties. The whole of this procedure is the EIA.

There is no publicly available written analysis of the process leading up to the decision, unlike the IPPC and Waste Licensing processes. In these, an inspector's report to the Agency, containing an analysis of the issues, is available for public comment, in the course of an objection. Essentially through, in each of these three processes it is the presentation of a report and then public comment.

The EPA by contrast in its "Guidelines on the Information to be Contained in Environmental Impact Statements"(P.7), lists public participation as one of its four basic tenets of EIA. In the same guide (para 2.3.7) it suggests, that apart from its basic statutory rights, the public could participate in the EIA process by making observations or submissions, in response to a preliminary invitation to participate in the procedural stages of scoping; evaluation of alternatives, and document review. Public involvement early in the process enables applicants to include or address concerns before it is too late.

It further asserts that, direct participation, i.e. face to face, communication is more likely to result in accurate and focused communication. In Para 2.3.6 the role of NGO's is discussed, and their input encouraged at the earliest stage possible.

"Early, open and constructive engagement has frequently proven to be beneficial to both the protection of the environment and to the quality of development projects", it states. However, at present these are recommendations only, and it is the proponents decision whether or not to involve the public, apart from the review of the final EIS.

The EIA Process

The Definition: "EIA can be defined as: the process of identifying, protecting evaluating and mitigating the biophysical, social and other relevant effects of development proposals, prior to major decisions being taken and commitments made" (Saddler, B. et al, 1999) The World Bank expands this definition and describes EIA as a procedure that; "evaluates a projects potential environmental risks and impacts in its area of influence; examines project alternatives; identifies ways of improving project selection, siting, planning, design, and implementation, by preventing, minimizing, mitigating or compensating for adverse environmental impacts and enhancing positive impacts" (Klees, R. 2002).

The main aim of the process is to stimulate thinking, and encourage action, and not just the ticking off of boxes just to get another report (Verheem, R. 2002).

The Principles of EIA

The IEA (Saddler, B. et al., 1999) identifies two sets of principles, 'basic' and 'operative'. The basic set applies to all stages of EIA as well as to Strategic Environment Assessment (SEA) of policies, plans and programmes. One of these states that the EIA process should be participative; providing appropriate opportunities to inform and involve the interested and affected public, and, that their inputs and concerns should be explicitly addressed in the documentation and decisionmaking" (Saddler, B. et al., 1999)

In a review of 25 years of National Environmental Policy Act (NEPA) the Council on Environmental Quality (CEQ) (CEQ 1997) stated "the success of a NEPA process heavily depends on whether an agency has systematically reached out to those who will be most affected by a proposal, gathered information and ideas from them, and responded to the input by modifying the proposal or adding alternatives, through the entire course of the planning process".

The EIA Cycle and Public Involvement

The EIA process should be applied as early as possible in decision making for a project, providing for the involvement and input of communities and industries affected, as well as the interested public (Saddler, B. et al., 1999)

The EIA cycle shown in Fig. 4 (above) as part of the planning system and project cycle should provide for the following: (European Commission, 2000).

Screening

Screening determines whether or not a proposal should be subject to EIA, and if so, at what level of detail. Consultation with potentially effected parties here will improve understanding of the nature and significance of potential impacts using 'local' knowledge, together with that of 'experts'.

The objectives here should be: to obtain a complete understanding of how the issue is viewed by all the significant stakeholders; and to identify the future levels of interest in public participation activities on this issue. Techniques could be used here from levels 1 and 2 of the participation tools list. (Eccleston, C.H. 2000. p.65) has a checklist of tasks to consider as part of the prescoping phase. Added to this should be the preparation of a public participation plan, which includes a publicly available record of the process. The use of exploratory thinking techniques such as brainstorming and snow-storming at this phase helps stimulate the creativity need to seek alternatives.

Scoping

This is a very early exercise in an EIA in which an attempt is made to identify the attributes of the components of the environment for which there is public (and professional) concerns and upon which the EIA should be focused (Singleton, R., Castle, P. and Short, D. 1999, p.103). Good scoping is the key to a successful EIA process. Public involvement here ensures that all the significant issues are identified, local knowledge about the area is incorporated, and alternatives are identified and considered. The principal objectives of public scoping are: (Eccleston, C.H. 2000, p.72)

- Identify public concerns and the expertise needed to investigate same.
- Identify alternatives to be examined
- Identify significant issues that need to be analysed, eliminating the unimportant.
- Identify problems and potential solution early in the process.
- Identify problems with the participation process and address same.
- Ensure that both the positive and negative aspects of the proposal are identified and studied.

- Identify potential mitigation measures.

A scoping information package should be put together to promote public involvement and to inform the scoping delegates. It should include:

- An invitation to participate showing how, when and where.
- A brief description of the working of the EIS process and the opportunities for participation.
- A description of the proposal, such that the objectives of the proponent are clearly laid out with maps, diagrams, figures etc.
- A description of the known potential impacts. A public notice should then be issued using any of the methods deemed necessary from Level 1 (Section 1).

As was mentioned in the section on P.P. tools, public meetings, whilst commonly used are not the best method, unless they are based on small group sessions and workshops. Public opinion surveys, citizen advisory committees or any other methods listed in Section No. 2 Simple methods may suffice for describing, synthenisizing and communicating information on the preproject environment and the potential impacts e.g. using checklists, matrices and networks. (Eccleston, C.H. 2000. p.75) (Australian EIA Network, 1996) gives a check list for planning a public scoping meeting, and warns that "outside entities that participate in scoping generally do so because they are opposed to the proposal; the remaining participants who support the proposal often do so because they stand to gain from it.

Not surprisingly numerous public scoping efforts have dissolved into sessions of frustration, dissension or outright confrontation. So it is essential to plan well and use professional facilitators or neutral moderators where debates are likely to be heated. Alongside the general public scoping, it can be of value to conduct focus group/workshop meetings with stakeholders having special interests or expertise, in order to examine more detailed or complex issues. The results of these meetings should be made public, and so become part of the general scoping process.

Eccleston, C.H. (2000 p.249) gives a methodology for handling the potentially voluminous, or complex scoping input, followed by an internal 'responsible authority' scoping session. This can consider issues including technical ones, using in-house expertise, and in greater depth than might otherwise be possible in a public setting.

The body conducting the EIA process, together with stakeholder representatives should then produce a post scoping document, summarising the scoping process, its findings and the reasons for decisions taken in reaching the scope for the resultant EIS Implementation Plans.

Impact Assessment/Mitigation

"Public involvement here can serve to ensure that the analysis and mitigation, necessary to avoid, minimize or offset predicted adverse impacts is relevant to local concerns, and accurately reflects local value and preferences". (European Commission, 2000).

Both impact mitigation, and the evaluation of significance are complex and time consuming processes, which cannot be dealt with by large groups. It is necessary, therefore, to use stakeholder representatives to work with the experts, reviewing the finding of the experts in workshop sessions or other facilitated small groups. Regular reporting of progress in the process, to the wider publics will maintain their 'ownership' and sense of involvement.

The Evaluation of Significance

This is a difficult issue to resolve even between specialists. It is, however, essential that the public be involved and their perspective included, as the interpretation of significance occupies a fluid boundary between science and politics (Australian EIA Network 1996).

The Canadian EA system of impact assessment is based solely on scientific, credible technical and other relevant information (C.E.A.A., 2001. The resulting determination of significance must be 'objective' and reasonable so as to withstand court challenge. By comparison in the US system, under NEPA, public opinion and the controversiality of the proposal help to identify and determine significance. Ultimately, however, the relevant authority will decide, taking all inputs on board.

Evaluation of significance is subjective, contingent on values and dependent on the environmental and community context (Australian EIA Network, 1996). The intrusion of wider public concerns and social values, into the significances evaluated by scientists is inevitable, as discussed elsewhere. The challenging nature of this part of the process is, therefore, one in which the use of facilitated small groups would be essential for the more contentious issues. More information, on the technical aspects of the evaluation of impacts. is given by Lein, K.L. (2003).

Preparation of EIS

It is essential to document clearly and impartially the impacts of the proposal, as well as those of the identified alternatives, the proposed mitigation methods, the significant effects and the concerns expressed by the public and communities affected by the proposal, as well as how those concerns were addressed.

The draft EIS, following internal review by the relevant body, should be made available for public review, and should include a non-technical summary. Review of the EIS The review of the EIS should determine whether the report; meets its terms of reference, provides a satisfactory assessment of the proposal(s); and contains the information required for decision making. Involvement of the public can ensure the quality and comprehensiveness of the assessment and help to reduce any bias in the analysis.

Before the public review, the EIS should be complete in every aspect, except this final opportunity for public feedback. Any major changes resulting from this public review would probably trigger a second public review, though good public participation early on should make this unlikely.

Proposals for electronic transmission of the document along with public access through libraries, site offices, public displays, information repositories, and presentations, should give wide availability. Methods of feedback should be widely publicised. One method could be an interactive web-site page for this purpose. All those who made major contributions to the process should receive a copy (finance permitting) or have access to a copy, along with the statutory bodies and the proponent. The draft should be available and publicised at least 15 days before any stakeholder meetings to discuss it.

The final EIS should then include the changes based on the relevant feedback in the draft, together with the responsible body's reasons for not including any other issues raised. Eccleston, C.H. (2000) gives useful guidelines on dealing with this stage of the process.

Decision Making

It is generally the role of the regulatory authority to approve or reject the proposal and, if it is approved, to establish the terms and conditions of its implementation, taking into account the EIS, and its public input.

Follow-up, Implementation and Monitoring

The participation of local representatives and NGO's in monitoring the operational impacts of a project can lead to the early identification of problems, and can foster a sense of public partnership.

Continual assessment of the EIA process

Continuous assessment of the process itself can ensure the implementation of the public participation plan, provide public transparency and strengthen the effectiveness of future EIA's.

EIA Best Practice

In its 'Directory of Impact Assessment Guidelines' (Donnelly, A., Dalal-Clayton, B., Hughes, R., 1998) the IIED (International Institute for Environment and Development) concludes that "greater attention to stakeholder involvement leads to better environmental assessment, and thus to the formulation of projects that deliver more social benefits, fewer environmental costs and greater economic and financial benefits". In looking at best practice world wide, it will not be possible to make assessments of their effectiveness within the scope of this study. However, it finds that in reality EIA guidelines and practice generally pay only lip service to real stakeholder involvement.

This is confirmed by the European and Central Asia Environmentally and Socially Sustainable Development Department (ECSSD) regional study of Eastern European and Central Asian Countries (Klees, R. 2002) where quite often, inclusive legislation is present but the reality of implementation is quite different. Constructive stakeholder involvement in EIA will rarely occur spontaneously. A proactive approach is required; starting with clear guidelines, legally enforceable by inexpensive court action; and financial assistance for stakeholders where significant costs would be incurred in order to take part in dialogue.

There are a number of good examples of proactive legislations and stakeholder support systems. Canada. Public participation in the federal assessment process is an implicit part of the Canadian Environmental Assessment Act, 1995, which was itself the result of five years nationwide consultation. The Canadian Environmental Assessment Agency (CEAA) was established to administer the Act and to ensure that opportunities for public participation are provided (C.E.A.A., 2001). The CEAA must reach out to the public for their input into screenings, comprehensive studies, mediations and panel reviews.

A public registry is established to ensure public access to records, and the responsible authority must address the need for a follow up programme to verify the accuracy of the assessment and efficiency of the mitigation measures.

A participant-funding programme, designed to provide limited funding to interested individuals and groups, is open both to those for and against a project. Its purpose is to enable effective participation at key stages of mediations and panel review.

The 'Guidelines on Environmental Assessment and Traditional Knowledge of Indigenous Peoples' (Donnelly, A. et al., 1998. p.29), formulated in Canada, draws attention to the limitations of short term scientific procedures to collect reliable and adequate information on which to base predictions. They also explore the very different perceptions of environment and development held by indigenous and non-indigenous stakeholders. New Zealand.

The Resource Management Act 1991 (RMA) states the two key principles: that decisions on environmental matters are most appropriately made by the communities most affected by those decisions, and that community participation is vital to effective resource management. This legislation is based in part on the Maori 'Kaitia Kitanga' (guardianship) of the 'Mauri' (life essence binding the physical and spiritual elements of all life). It is the first attempt world wide at an holistic legislative approach to the management of environmental resources (Ministry for the Environment, New Zealand, 2003). The Act simplifies access for the public by putting all activities that impinge on environmental resources under the same resource consent procedures. A resource consent gives a person or organization permission to sue or develop a natural or physical resource, and/or carry out an activity that affects the environment in some way for a stated period. Resource consent replace all the many different permissions granted under previous law. Its simple appeals structure, Environmental Legal Assistance Fund, and its insistence that decision makers actively consider alternatives to regulation in preparing policy statements and plans (Section 32 of the RMA), makes the RMA a highly accessible process for the public to become involved with. The Arctic.

The Arctic Environmental Protection Strategy (AEPS) agreed by the eight Arctic countries lays great importance on the integration of two ways of knowing: Traditional Indigenous Knowledge and Scientific Knowledge (A.E.P.S., 2002). Experience gained in mapping traditional knowledge on the Belluga whale in ten indigenous communities on the Alaska and Russian sides of the Bering Sea, have lead to the inclusion of the use of traditional knowledge alongside general public participation as two of the important features of the "Guidelines for EIS in the Artic"(A.E.P.S., 2002)

The USA. The Council on Environmental Quality (CEQ) has been central in strengthening procedures and guidelines in the USA since the National Environmental Protection Act 1969 (NEPA) was enacted. Public scoping of all federal agency projects requiring an EIA is the norm. Publicity in various media is combined with invitations for comments or suggestions, which can be submitted via the post, at various public open houses or public meetings, and via e-mail. Sri Lanka.

The scoping process here involves formal and informal meetings with people who may be affected by the proposed project either directly or indirectly, or who may have special knowledge of the project area and its environment (Tille Kerante, L.S.G., 2003). Australia. The Environment

Protection and Bio Diversity Conservation Act 1999 (EPBC Act) establishes a Commonwealth process for environmental assessment and approval of proposed actions likely to have a significant impact. This is the culmination of over 25 years of EIA in Australia.

Five options are listed under Section 8 of the EPBC Act (Environment Australia, 2003). The first three, preliminary documentation, public environment report (PER) and environmental impact statement involve the following steps: proponent prepares and publishes draft document; public comment period; proponent finalises document incorporating comments; Secretary of Environment Australia reports to the Commonwealth Environment Minister who makes a decision.

All documents prepared by the proponent will follow guidelines (terms of reference) laid down by Environment Australia, and formulated by consultation with the relevant stakeholders. Where the impacts are likely to be high, and prior to preparation of the final EIS, or where greater public involvement is to be ensured, a Public Inquiry is held. All information leading up to the inquiry, including the reasons for having same, are made available. The hearings are all held in public and all submissions made available to the wider public. Quebec, Canada. (Australia EIA Network, 1996 p.25) Established in 1978, the Bureau d'Audiences Publiques Sur L'Environnement (BAPE) is a quasi-judicial advisory body with a strong record of public participation. The Minister on request from any person, group or municipality may instruct BAPE to hold; a fact finding enquiry; carry out a combined enquiry and mediation; or undertake public hearings. Its role is to enquire into any question, relating to the quality of the environment, that has been submitted to it by the Minister of the Environment and report back on its findings. It is not like An Bord Pleanala, which is an appeals board only. B.A.P.E. once instructed, plays a proactive role.

A BAPE Commissioner meets with the parties and mediates to seek a joint solution. Failing this, two to five Commissioners serve as an independent review board. Initially, the public has 45 days to review an EIA and request mediation or a hearing. BAPE must report to the Minister, within 120 days of receiving a request for a public hearing or 60 days from receiving a request for an inquiry or inquiry with mediation. Success rates of mediation by BAPE are 75%. The option for a further period for a hearing helps circumvent the feeling of negotiating under time pressure. At the same time, proponents know that these timescales are absolute.

The mediation process provides an effective and efficient means of resolving conflict with a 75% success rate in 1978. Institutional arrangements for EIA Review (Australian EIA Network (1996). Permanent Commissioners in Holland (Netherlands Commission for EIA, 2002) and Italy review the quality of the EIS. The Dutch Commission for EIA remains outside political decision making, it's role is to give an expert (selecting from a panel of 400) opinion on the quality and correctness of an EIS. The Italian Commission by contrast can make a judgment on the environmental compatibility of a project, as well as the quality of its EIS. In Canada, for certain major projects (50 in the period 1975-1995), ad hoc review panels, or inquiry bodies are used. A panel's major responsibility is to establish terms of reference (TORs) for an EIS and to conduct a public review of its suitability. The Parliamentary Commissioner on the Environment in New Zealand can also use panel reviews.

Ireland, Planning Major Roads. Although the Roads Act, 1993 as amended by the Planning and Development Act, 2000 only specify public consultations in the case of compulsory purchase orders (CPO's) or post EIS publication, the NRA in its 'National Roads Project Management Guidelines' sees public consultation as a vitally important and integral part of the various phases of major project planning. The first round of public consultation occurs in the preliminary planning stages as part of the feasibility/ constraints study. Following the publication of the Consultants Report, which is based on the public submission and the consultants report, the possible route options and their general impacts are identified and the public consulted again. This leads to the route selection, and only then, is the EIS drafted. (NRA, 2000).

Research Into EIA. The EC Joint Research Centre (ISPRA) is at present conducting a study on behalf of the EC Environment DG into EIA and SEA research. One of the six areas for research relating to EIA's is public participation (Europa, 2003). A more general research 'Programme on Environmental Decision Making' (2001-2006) (E.S.R.C. 2001) aims to investigate and elucidate a set of criteria for better environmental decision. Further examples of good practice in public involvement in EIA and other areas of Environmental governance can be accessed through the good practices suite of the United Nations web-site: Virtual Conference for Integrating Environmental Considerations into Economic Policy Making Processes (ESCAP).

Strategic Environmental Assessment (SEA)

"SEA is a systematic process for evaluating the environmental consequences of proposed policy, plan or programme initiatives, in order to ensure they are fully included and appropriately addressed at the earliest appropriate stage of decision making on part with economic and social considerations". (Sadler and Verheem 1996). In their principles for SEA, Dalal-Clayton and Sadlier(Donnelly, A. et al, 1998, pp 31-42) call for transparency and openness, including early consultation, availability of comprehensible documents and explanation of how all decisions are arrived at.

The development of a pro-active SEA process is a subject too big for this study, but is one in which public participation should play a part as large as that proposed for the EIA process. The Dutch systems for EIA and SEA are identical in structure, and in both public involvement at the scoping phase is required. (Netherlands Commission for EIA 2002). By contrast, the new European Directive leaves consultation until much later in the process. 'Directive 2001/42/EC (The SEA Directive) (117) states that prior to the adoption of a plan or programme, or its submission to the legislative process, the competent authority must carry out an environmental report.

The draft plan or programme and the environmental report must be made available to the relevant authorities and to the public, who will be able to express their views on same, prior to it's adoption or submission to the legislative process. The relevant authority, following where necessary trans-boundary consultations, will inform all the parties concerned which have been consulted of its decision. With the decision will be a statement summarizing amongst others the results of the consultations, and the reasons for choosing the final plan or programme.

THE SURVEY

The purpose of the survey was to make an assessment of the present state of public participation in EDM in Ireland. An attempt as made to look at:

- o Attitudes of the main stakeholders.
- o Their experiences with public participation.
- o Present in-house capacity of regulators and industry to enable public participation.
- o In-house capacity building measures.
- o Capacity building in the wider community.

Methodology

The questionnaire was based on the results of the literature survey, personal experience and feedback from trial questionnaires. The survey was conducted by e-mail, except the industry section, which due to confidentiality issues had to be posted by the college administrator. Where appropriate, the same questions were asked of all stakeholders, but specific questions relevant to each group were also included.

The public questionnaires included a preliminary set of open-ended questions. These were designed to establish the person's particular experiences and involvement in dealing with a particular EDM process.

Industry-15 Completed Questionnaires.

Postal contact was made with 40 graduates of the H.Dip in Environmental Protection. The responses came mainly from Environmental Engineers and Environmental Safety Officers, together with one consultant and one company CEO.

The industries involved included:

- Wind energy Distillation
- Pharmacies Biological
- Metallurgy Galvanizing
- Chemical Coating Turbines
- Aircraft Man-made fibres

Local Authority, Environmental Sections-12 completed questionnaires.

Questionnaires were circulated to all 34 local authorities, some of which did not have separate environment sections. A 35% return rate.

Environmental Awareness Officers (EAOs)-16 completed questionnaires. A 47% return rate.

Although based in local authorities, the function of EAOs is separate.

Department Of The Environment And Local Government -1 completed questionnaire.

This return represents both the individual opinions of the respondent and department policy where appropriate

Other Relevant Departments and Agencies-2 completed questionnaires.

One was from a Senior Environment Officer with a Fisheries Board, the other the Coordinator of a River Basin District (RBM).

DISCUSSION OF RESULTS

In this section the results, are summarized, and then their significance discussed.

The experiences of the Public Half of the 20 Public respondents became involved in EDM processes as a result of particular issues that arose in their communities, and they then attempted to engage with the relevant regulatory authority. The other half became involved with longer-term issues, also in their communities, but which required on-going involvement over long periods, up to five years in one case.

Whilst 40% saw the outcome of their involvement as positive, and all but one saw it as a worthwhile learning exercise, the overall experience of dealing with proponents and regulators has generally been negative. Despite starting from a broadly constructive point of view, these respondents ended up largely cynical, frustrated and angry.

Only 17% of respondents believed the process in which they were involved to be legitimate and consistent with the aim of environmental protection. Their trust in the various regulatory agencies, following their experiences, was also well below acceptable levels, with zero for the EPA. From Question A.11, it can be seen that 75% of the respondents had experience of the planning process, but only 35-40% had experience of the other processes. Some of the reasons for these opinions can be seen from the responses to other questions.

There was a broad desire expressed for proper consultation. These desires were met with tight time constraints and minimal help in understanding the process involved.

In some cases the treatment received was a hindrance in trying to get to grips with procedures. Only thirty percent felt they were being listened to, and only 37% had the reasons for the final decision explained to them. According to the respondents, none of the EIS studied by the respondents involved fulfilled the basic guidelines of the EPA, and no public post-project monitoring was carried out.

38% felt that the responsible body was competent to enforce the decision made. Only 43-60 % of respondents, depending on information source, felt access to documents was easy, despite the relevant legislation.

Add to this that, at the out set, only 35% understood the process they were engaging in, let alone the intricacies of the issues involved; that very few found it easy to locate legal or technical expertise, and that paying for these meant that, for 45%, fund-raising became a major part of their campaign. Individual expenditures were huge, ranging from □70 to □20,000 in one year, extending in some cases over many years, coupled with time spent of 70 hours total, to 15-20 hours per week again over years.

So what is the overall scenario emerging?

Here we have twenty individuals, most of whom are linked to groups with shared values, who attempt, as they see it, to protect their local environment, whilst perhaps trying to affect the bigger picture.

Apart from one laudable exception, it seems that they were not greeted with open arms. What should have been a positive first entry for many into the processes of EDM has instead left them cynical and angry, with one respondent opting out for the future, calling instead for nuisancemaking, street politics and media events, in order to create political pressure.

Here was a respondent who, in his words, started with optimism and campaigned constructively, attempting to engage with the regulators and proponents alike, and who found this engagement futile. The rest of the respondents were not far from sharing this opinion.

Finally, from Question 18, it is clear that the quality of EISs, that accompanied the EDM processes experienced by the respondents, was generally perceived to be of a low standard.

CAPACITY FOR PUBLIC PARTICIPATION

From the responses to B6 and B7 it is clear that 60% of the industry do not have formal lines of communication with the wider community or help with comprehension of technical documents, and are generally passive in this regard.

By contrast, Local Authority (LA) staff do see a strong role for themselves in providing assistance, and in understanding legislation and technical documents, even though only 58% have clearly laid down procedures for requests for information. Even then the procedures are generally only in relation to requests under the Freedom of Information Act 1997. Only 27% of LA respondents have a freephone call-line, compared to 40% in industry. LA staff assess their own capacity, to give good access to information on the environment, very highly. (B11) Looking at B8 and B9 it is clear that the council staff are inclined towards the more traditional, generally passive methods of information giving but with a strong reliance on website development.

EAOs by contrast and by the nature of their job, are more pro-active and use a wider range of tools. Many local authorities are still developing their websites, and so the availability of information on the different LA websites varies dramatically in quality and quantity, as can be seen from responses to B10 and B12.

B14 responses showed a generally low public access skills base, with not surprisingly, EAOs being better trained overall.

The sources of information were used by all the stakeholders was the subject of B16. The Public generally used independent sources, although surprisingly few used the excellent resources of ENFO.

The use of the local authority, as a source, received a rating of 20% of the public total. Council staff also made little use of ENFO, sourcing most of their information with the EPA and the DOELG with some use of consultants. By contrast then ENFO was the first choice of EAOs, with the EPA and DOELG second.

Reflected here may be the lack of trust felt by the public in the regulatory authorities. It should be noted that the DOELG was not offered to the public stakeholders as a choice. The Local Authority is a necessary source of some information for most EDM processes and so received a larger number of individual public responses than the other sources, but its average score was 2.4 compared with 4.8 for "other" and 3.8 for "experts".

To summarise, the capacity to provide good access to public participation is at best patchy, despite the response by local authorities, to B11, asserting c.90% capacity to give good access to information. Whilst the inter-net is an invaluable public access tool for public bodies and industries, the existence of the 'digital divide' is not mentioned by anyone. Most of the websites are still at an early stage of development, and very few are interactive. Yet the inter-net seems to be the main focus for the council respondents in particular.

Summary and Discussion of Results

Given the inevitable drive to build a sustainable society in which the economic sociologic and environmental aspects of development are all considered, the importance of including all the stakeholders in environmental decision-making (EDM) becomes increasingly apparent. All actions for change have consequences, and the moves to greater public participation have their costs as well as benefits.

It is necessary for all people to decide whether these costs are greater or lesser than the costs of no change. In order to make that choice it is essential to have as much information as possible. This study attempted to draw together some of the arguments for and against that change. Making the assumption that some changes are coming already, as a result of the Aarhus Convention and its offspring, this study has examined ways of making public participation in EDM effective.

In understanding any change, it is necessary to know where you are starting from, and the possible routes that might be traveled to get there. To this end a survey was made of the major stakeholders in EDM processes, industry, the regulators, and the public, in order to see the starting point, as well as some of the possible ways forward.

In the time-scale of this treatise, it was not possible to get statistically representative samples of either industry or public opinion, although the local authority sample was sizeable. 35% of environment departments, 47% of EAO's covered in total 65% of Local Authorities.

The methods by which the public and industry groups were contacted were very specific, so this and their self-selection in responding should be taken into account in interpreting the results. Every ENGO on the ENFO list was surveyed, but only one member of one organization responded, despite each receiving two reminders. Several said they did not have the time.

The authors impression was that they mostly saw public participation as an irrelevance to them. The IFA (Irish Farmer's Association), I.B.E.C. (Irish Business and Employers Confederation) and all the government departments with an obvious environmental remit were also surveyed to no avail, apart from the DOELG, who did respond.

From this the conclusion could be drawn that either these bodies are not interested in the survey, or they are not interested in public participation, or they see participatory practices as a dilution of their power. This is an area for future research.

In order to get a reasonable return on the survey, it was necessary to send out the questionnaire long before the literature survey was completed. This meant that some questions that should have been asked were not, and others were asked badly. An example of the latter would be Question B.1.-Should the public be excluded from any particular area of EDM? The strong response against exclusion does not, however, indicate whether they think the public should be included. With regard to the former, it would have been helpful to find out why each group thought public participation should entail.

Taking the above reservations into account, and in the context of the literature survey, a number of indicators for future action can be taken from the results of the questionnaire survey, as will be seen. The questionnaire looked in turn at the experiences and opinions of the stakeholders, as well as their capacity to deliver public access to information and EDM. Opinions on capacity building measures were also sought.

Putting together the experiences of the different stakeholders, it fast becomes apparent that they have been very different.

On the one hand, we had the public stakeholders, who clearly felt that they have been largely ignored despite their best attempts at constructive engagement with the regulatory authorities. Their commitment cannot be denied in both time and money, and yet they seem to have met, in general, with unhelpful and sometimes obstructive treatment by public servants at local and national level. On the other hand we heard from those people working in Local Authorities and the DOELG that they were confident in their ability to provide good information to the public, and that although it might slow EDM processes down, they had the majority opinion that public participation had many positive attributes. They also felt that their staff were enthusiastic (c65%) though apprehensive about public involvement in EDM. Industry respondents were even more enthusiastic for public involvement, though they did not have the same strength of belief in its ability to reduce conflict or improve project design.

Clearly something is wrong here. Assuming that all the respondents are genuine in their responses, and that it would appear that they would all wish to operate in a more open way, with the protection of the environment as an end point, there is clearly either a communications breakdown, or else the structures and laws set up to protect the environment are unable to deliver public participation.

It seems like a mixture of the two. In order to communicate well, and under time pressure, it is necessary to speak the same language and to know the rules of engagement. It seems that one third of the public did not understand the EDM processes they were entering into, and only two thirds understood at the end, with very few (27%) getting any assistance along the way. And yet the public servants seem more than willing some (c80%). They were also willing to help with technical matters (c75%).

Independent expertise was hard got by the public and expensive, and yet in situations where the proponent of the project or plan was a government body, they saw themselves as paying for expertise on both sides of the debate.

What happens when a member of the public requests information? Only 58% of Local Authorities said they had strictly laid down procedures and only 40% of industry. Where guaranteed call back hotlines operate (only 25% of Local Authorities) response times were from twenty minutes to 72 hours.

It would be useful here to know how the front-of-house people in Local Authorities feel about their ability to deal with these questions. Do they have the training, back-up and staff numbers enable them to give the service they would like. These questions, unfortunately, were not put. However, in looking at questions relating to capacity, the following becomes clear. For enabling information delivery, the web-site is a powerful tool but limited in its accessibility, and yet it seems to be the focus of most of the council respondents. At the same time, despite some shining stars, environmental information on Local Authority web-sites, is still fairly restricted, as in that on the E.P.A. and DOELG web-sites. However, they are all in the process of improving, and in the case of the latter two, undergoing radical overhauls. Links from Local Authority web110 sites to other sources of information ranged from 50% for local bye-laws, down to 29% for European Directives.

The author is not aware of any research into who accesses the existing sites, or who knows that they exist, and yet they seem to be the main focus for government resources, and local government officials, in terms of improving accessibility.

Whilst EAO's, whose job is to educate and interact with the public on a daily basis, rate their public access skills highly, they are not the ones on the front line in EDM processes. The percentages of Industry respondents in particular and Council respondents that replied positively to having public access skills was low in general, except in media skills, and for the council officials with regard to public meetings.

It is not to be expected that these two groups would be highly trained in these areas, it is not their first priority, but these skills, particularly the facilitators skills needed to work with small groups and in workshops, should be available, in house, in every government agency or Local Authority dealing with the public in EDM.

These skills would also be useful in other aspects of their internal and external business, e.g. Partnership and the development of E.M.A.S. Generally though, when asked about proactive methods of giving information and receiving feedback, industry was very passive in its approach. Council officials also tended to be passive but showed a greater willingness to engage with the public. As mentioned earlier, the web-site seemed to be the tool of choice for council officials.

The EAO's and the DOELG were both very proactive in information giving and feedback but the EAO's as educators showed the widest range of skills.

Access to outside sources of information is another area of capacity that needs examination. Question B.16. showed council officials, who deal largely with the technical/decision making aspects of EDM, opting for sources of information that enable them in these tasks. Similarly EAO's, who used the educational resources of ENFO also used the DOELG and EPA.

By contrast, the public largely used independent sources which got very high average scores of 4.8 111 for 'other' and 3.8 for 'experts' compared with only 2.4 (out of 5) for the Local Authority, even though 75% of the respondents were involved in EDM processes to do with Local Authorities and planning. None of the public respondents were from Dublin, which might explain the low use of ENFO (8%). Those that did, gave it a score of 3.2 on average. The author would score ENFO much higher as a resource, from his own experience, and would recommend that ENFO be resourced such that it can make its presence felt in every region of the country.

The use of independent sources of information by the public would tie in with Question A.11 where only one third of the 75% of respondents dealing with local government planning issues had increased trust at the processes end, and a lower figure dealing with waste permitting (22%). Trust in the EPA was zero and in An Bord Pleanala 43%.

The only area of actual EDM discussed in the questionnaire was the EIA process. Clearly the public respondents that had had experience of EIS's found them of very poor quality indeed. Alongside this, we have the Industry respondents, who were 73% enthusiastic about public involvement in EDM, giving only 57% support for greater public involvement in the EIA process, with 53% agreeing with public involvement in the 'comparison of alternatives'.

So this is the base line, we had a largely disgruntled public, an enthusiastic but under-skilled and under-resourced public service, and an industry response that was generally lukewarm. How did these same respondents see the way forward, what capacity building did they see as useful? When asked what changes in their organization would improve public access to environmental information and EDM, Industry respondents look to increase the transparency of their enterprises, and showed a desire to reach out into the community.

Local Authority and EAO respondents both looked for increased resources, staff training and improved web-sites. Regular meetings with key stakeholders in the community, coupled with increased public access to Strategic Policy Committees were both creative suggestions.

When asked to make choices about specific capacity building measures, Council respondents opted for staff training in public access skills and increased access to information for their staff; whilst the main choices for industry were staff training in legislation and technical details alongside training in public access skills. EAO's looked for advertising budgets to assist with their education programmes.

The penultimate question, B.19, dealt with capacity building measures for the wider community,.

The measures receiving the highest cumulative scores were; 'collaborative participation where all stakeholders share information and decision-making' at 60%, and 'an environmental ombudsman' at 58%.

Apart from this, Industry and the Public both scored very highly for the three interventionist measures of; an expert panel to resolve disputes over 'facts; a community based independent environmental information broker, to give that expertise sought by the public; and an ombudsman to act as a final arbiter if all else fails.

The final question B.20, asked of all, but the public, looked for advice to others, based on the experiences of the respondents. Most of the responses were positive to PP, suggesting a more personal approach to the EDM processes with PP incorporated into the EDM process at the earliest possible stage.

Two responses from Industry felt the politico/sociological climate in Ireland was not mature enough. Perhaps by reaching out into the communities, of which they are but one part, they might be surprised. There is a great wealth of detail in the responses of all the stakeholders, which can be seen in Appendix II and which in its variety emphasizes the fact that, as in all human activity, nothing is black or white.

CONCLUSIONS

It is clear that the present situation with legislation, that largely restricts public participation to 'report and comment' activities, is likely to leave the public stakeholders with feelings of frustration and alienation.

It is also clear, that without the necessary changes in practice, and increases in resources for skills training and more personnel, public servants cannot deliver effective access to the public.

The lack of trust shown by the public in the regulatory authorities, put together with the apprehensions of the regulators and industry, about public involvement in EDM, will require a great deal of work to change. Open and inclusive debate, amongst all stakeholders, about what changes are needed in their inter-relationships and in the distribution of decision-making power are essential.

The Directives coming from Europe and emanating from the Aarhus Convention should be seen as only the minimum standards, and the spirit of Principle 10 brought in when they are enacted into Irish law. Indeed, the drafting of new legislation should itself be open to public consultation.

Greater public awareness of their rights to be involved in EDM processes, and greater knowledge of how to be involved would, it was agreed, lead to the end-game of greater environmental protection. The more the public are involved in EDM, the better educated they will become on environmental issues, and hopefully then the safer the environment will be from thoughtless plundering and degradation of natural resources.

The way forward then lies in; greater openness, a willingness, on the part of regulators and industry, to reach out pro-actively into their communities, a detailed assessment of the capacity needed to deliver participation, and a commitment on behalf of all stakeholders to work together for their mutual benefit, and a sustainable future for all life in this Biosphere of which we are a small but potentially very destructive part.

RECOMMENDATIONS

- That Ireland, as a signatory, actively seeks to carry the spirit of Principle 10 into all areas of EDM.
- That Irish Governments, in drafting legislation on the environment, proactively consult all stakeholders beforehand.
- That the Irish Government finance an assessment of Principle 10 implementation, as agreed under Partnership for Principle 10, using an independent assessment method such as the Access Initiative. In this way, capacity building will have a base-line to focus on.
- That regulators proactively seek out stakeholders and encourage their involvement in EDM well before the final decision is made e.g. calling door to door in the neighbourhood of a proposed development.
- That timescales for submissions about EDM should be more flexible, allowing more time for the consideration of complex issues.
- That funding be made available, to enable public participation, in the form of grants for legal and technical assistance.
- That the public be involved as early as possible in the EIA process. Preferably from the screening phase onwards, sharing in the information flow that leads to the EIS.
- That reasons be publicly given for the final decisions in all EDM including the decisions of An Bord Pleanala, and that responses to submissions are included with same.
- That the role of ENFO should be expanded both geographically and in terms of its service provision. This latter to include an independent environmental information broker, with access to a panel of independent experts.

PUBLIC PARTICIPATION PRACTICE

Programmes to encourage the public in EDM processes often have multiple purposes, and no simple formula exists for success. In each case a programme has to be designed to fit the particular combination of process, proponent and public. A number of case studies are given by Acland, A. (2002. pps 25-30), Ortolano (2997, pps.409-417) and the European Commission (COM (2002) 704 final. pps 62-72) and are illustrative of this point. The proponent of an EDM process be it a public agency or private enterprise must be clear as to their aim in involving the public.There may also be different aims at different stages of a process, for example within the EIA process from the elicitation of values relevant to site selection at the project design stage, to the optimisation of trust and credibility at the monitoring stage (Petts, J. and Leach, B. 2000. p.20). In order to put together the aims with the methods it is necessary to have a plan:

Planning for Public Participation

In order to make a coherent plan it is necessary for the practitioner to establish the objectives of the process, based on the overall aims. It is then important to identify which public(s) the participatory process is trying to include.

Objective Setting

Clear objectives in planning a public participation process are essential, and ideally they are delineated by consultation with a representative group of stakeholders. A community assessment process, involving community interviews (Marxen, J. 2001. 6-4), would provide a good basis for a plan. The California Department of toxic Substance Control, in its 'remediation and permit procedures' involves a public participation specialist to review the findings of such interviews (Marxen, J. 2001. 6-20). It is equally essential that all those involved, both active and nonactive participants understand the objectives, as well as the limitations on what can be achieved. The connection between process and outcome must be clear to all. Ongoing process evaluation is essential to keep a clear focus on the objectives (Cuff, J. 2001. p.75). The two basic reasons for identifying and classifying objectives are; that objectives change over the various stages of a process, and that some participation techniques are better than others for achieving particular objectives (Canter, L.W. 1996. p.593).

Important Factors in Objective Setting

Apart from the above, there are a number of specific factors that underlie the setting of objectives, that will need to be considered at the outset, and will be present throughout the EDM process. Environmental Protection The main purpose of EDM, and thus the main purpose of public participation in EDM is to achieve protection, conservation, and wise management of the environment. This can only be achieved if the proponent properly collects (and acts upon) evidence, opinions and perspectives from all the interested or affected stakeholders, who are to be fully involved in the EDM process, and from the earliest opportunity. Legal Requirements Appendix 1 shows the major current legislation relating to public participation in EDM in Ireland.

Opportunities to participate are present throughout the process of making a Development Plan, a Waste Management Plan and a River Basin Management Plan, however, most of the other processes are of the 'report and comment' type where the public's reaction is sought to an already drafted document. As discussed elsewhere, this situation is changing with the Directives that are emanating from Europe implementing the Aarhus convention. The legal processes, their limitations and the concomitant responsibilities should be made clear to all the participants at the outset. Proponent Objectives vs Public Objectives. The various stakeholders in an EDM process can have very different expectations of that process. The proponent may be only conducting a public relations exercise or fulfilling a legal requirement, having a final decision already made. The participants may want an influential voice in the EDM process and expect their input to be valued and included.

The proponent, seeking a mandate for a particular project, may seek to defuse opposition through consensus building. Conversely, the individuals or groups involved may be largely uncompromising on their expected outcomes. The proponent may be restricted by time and/or money, whereas the public participants may be highly motivated by value judgments made on a particular issue, and prepared to commit much personal time and energy to it. Participation can then be viewed as a tool that the public and the project proponent or planmaking agency can use for their own purposes. Two Way Communication. The development of a two-way process of communication between the regulatory authority and the public is essential in order to identify public concerns and values; inform citizens of authority's plans; and inform the authority about alternatives and impacts.

Experts vs. Public. Accepting that the role of regulatory authorities is to serve the public interest, the challenge is to integrate technocratic and democratic contributions, when addressing complex issues (Lein, K.L. 2003. p.57). Dearing and Rogers (1996) state bluntly that, "scientific research results do not play a role in the agenda setting process". The social construction of reality is what gives issues their importance, such that the building of a waste incinerator creates much stronger reactions and media attention that regular back yard burning of waste, regardless of the relative levels of dioxin release. Levels of Participation. The setting of objectives is constrained by the level of participation being allowed in any one process. It is useful to look at "Arnstein's ladder" (Ortolano, L. 1997. p.417) /and for the agency or proponent to analyse its motivations in establishing the process.

1. Manipulation

2. Ther Non-Participation. These levels assume a passive apy audience, which is given information that may be partial or constructed.
3. Informing
4. Consultation Degrees of Tokenism People are told what is about to happen, what is happening now, or what has already happened.
5. Placation People have a voice, but no power.
6. Partnership People's views have a small influence on decisions made by traditional power holders.
7. Delegated Power Degrees of Citizen Power People negotiate with traditional power holders, agreeing rules, roles, responsibilities and levels of control.
8. Citizen Control Some power is delegated.

Arnstein represents the levels of citizen participation as the 8 rungs of a ladder and groups the rungs into three groups. The continuum stretches from going through the empty ritual of nonparticipation, to having the real power needed to affect the outcome of the process. Arnstein describes this first category non-participation as tactics whose real objective is "to enable power holders to educate, or cure the participants". In the degrees of tokenism Arnstein argues that when these are "proffered by power holders as the total extent of participation, citizens may indeed hear and be heard. But they lack the power to ensure that they are heeded".

The final category, citizen power, involves citizens-agency partnerships such as those in the Bantry Bay Coastal Zone Charter and finally programmes in which citizens are in control, or can veto agency decisions.

The further the process is from centralised power, the further up that ladder it is possible to go. Thus, at the European Union level, public participation has reached its peak at the level of consultation, whereas there are now many local partnerships under Agenda 21. Generally, however, regulatory authorities are not empowered to go beyond the consultative level in public participation. Wilcox, D. (1994) adapted Arnstein's Ladder to give four categories of participation methods, and these present an easier model for the purposes of this study, as will be seen later. They are:

1. Education and information provision
2. Information feedback
3. Involvement and consultation
4. Extended involvement

These categories are explained in Section 5.4. To enable the selecting of objectives, Canter, L.W. (1996) looks at a number of systems for categorising them. One such gives three general types of objectives, which further divide to give seven more specific, viz: Public Relations-Legitimizing the role of the proponent or regulatory body. Developing confidence and trust in the stakeholders. Information-Diagnosis of problems and needs. Development of alternative solutions. Evaluation of consequences and alternatives. Conflict resolution-Consensus seeking. Depolarizing interests. One international body, The Institute for Participatory Management and Planning (Connor, D.M. 2003) (I.P.M.P. 2003) goes much further in breaking down the objectives, with fifteen basic objectives under three categories.

The IPMP list is based on enabling a public participation practitioner, working for a regulatory authority or proponent for a project, to analyse their objectives, and then establish the best methods to achieve them. Each objective is coupled to a check-list. Viz:

- Responsibility. Establish the legitimacy of your Agency and you project.
- Maintain the legitimacy of your Agency and your project.

- Establish the legitimacy of your process.
- Maintain the legitimacy of your process.
- Establish and maintain the legitimacy of assumptions and earlier decisions.
- Responsiveness. Get to know all your potentially affected interests.
- Get to see the project through their eyes.
- Identify all potentially relevant problems.
- Generate solutions.
- Identify and classify the key issues.
- Effectiveness. Nurture and protect your credibility.
- Have your communication received and understood.
- Receive and understand information that's communicated to you.
- Search for common ground between polarizod, potentially affected interests.
- Mediate between polarized interests.
- It is clear from this that one of the main objectives of practitioners is to maintain their credibility and that of the process, without which all else is in vain.

The code of ethics for practitioners, from the International Association for Public Participation Practitioners (IAP2), is also a useful general guide in setting the objectives of an effective process (IAPZ. (a). 21/5/03).

- o Purpose. We support public participation as a process to make better decisions that incorporate the interests and concerns of all affected stake holders and meet the needs of the decision making body.
- o Role of the Practitioner. We will enhance the public's participation in the decision-making process and assist decision makers in being responsive to the public's concerns and suggestions.

- o Trust. We will undertake and encourage actions that built trust and credibility for the process among all participants.
- o Defining the Public's Role. We will carefully consider and accurately portray the public's role in the decision making process.
- o Openness. We will encourage the disclosure of information relevant to the public's understanding and evaluation of a decision.
- o Respect for Communities. We will avoid strategies that risk polarising community interests, or that appear to 'divide and conquer'.
- o Advocacy. We will advocate for the process, and not for any interest, party or outcome.
- o Commitments. We will ensure that all commitments made to the public, including those by the decision maker, are made in good faith.

Identifying the Public

There are as many publics as there are different people who care, positively or negatively, about a project. Who they are may depend on their ethical, moral, interest, welfare etc. viewpoints. Other reasons the public may be affected by regulatory decisions and plans include:

- o Proximity-Pollution, Property values, Employment.
- o Economics-Landowners, House-owners.
- o Use-Amenity value, rights of way, vista.
- o Social and Environmental Issues-Justice and Risk.
- o Values-Animal Rights, Ecology, Religion.

These members of the public could be described as 'stakeholders' alongside the proponent, regulatory bodies, industry and NGO's. Wates, N. (2000 p.15) defines stakeholders as "persons or organisations with an interest, because they will be effected and may have some influence."

Whatever kind of process is chosen, it is essential to define which members of the public are stakeholders (Yosie, T.F. and Herbst, T.D. 1998. p.19). Different stakeholders are affected to varying degrees, and the greater their interest in a project, or the greater the potential impact on them, then the more involvement they should have. There are also those who could affect or impact on the project.

An individual's interest may change over the course of the participation process, and so they may opt in or out at various stages, others may want to be involved throughout the process. Some may purposefully reject participation. It is essential, however, to ensure that no group of persons is excluded, as this, whether intentional or not, will be resented, and may lead to action being taken in the legal or political spheres. The number of potential public stakeholders on some issues is practically limitless, however, not everyone can sit around the same table. The challenge then is to balance the need to consider the many view points of all the stakeholders, with the practical considerations of convening a group of individuals who have a role in making, or directly influencing, decisions.

'Tiered' stakeholder involvement is one approach used to strike that balance, and in it three categories are defined:

- o Those who want to be directly involved in the process
- o Those affected who just wish to be kept informed
- o Anyone with an interest in the project

Many stakeholders are not able to effectively participate in deliberations even when they are the most potentially affected. This may be due to poor language skills, poor self-esteem, lower educational achievement, physical handicap or intellectual disability and so on. Many remain unaware of their opportunity to participate.

The environmental Justice Programme of the US EPA, attempts to address these issues requiring "the decision makers to seek out and facilitate the involvement of those potentially affected". (US. E.P.A., 17/11/02). Participation specialists distinguish three ways of identifying segments of the public (Ortolano, L. 1997. p.405).

- o Self identification
- o Staff identification
- o Third party identification

In Self Identification, individuals and groups come forward as a result of publicity by the agency or body conducting the participation programme. These are "active" responses and should be included in a mailing list and kept continually informed of the process. Staff identification-regulatory agency personnel learn from experience the names of individuals and groups that might be interested in involvement. GIS (Global Information Systems) are being used increasingly by agencies to enable both the public and themselves to ascertain the locality of environmental problems, as well as that of the stakeholders geographically relevant to them (US, E.P.A. 2000). Other staff identification techniques include: Electoral Registers; Mailing Lists; analysis of local newspaper stories; local organisation lists; and maps. Third party identification can be achieved by asking those already involved about other individuals or groups that should be sought out. Local officials in other agencies can be asked to suggest people who may want to be involved. As each person becomes involved, they are asked to suggest others, until the response becomes minimal.

Various public involvement groups check lists exist, and can be useful as another way to ensure that no group is overlooked. (Canter, L.W. 1996. p.598) (DOELG, 1995. p.16) US E.P.A. 1999. 2-1). But it is particularly important to bear in mind that in some areas of cities, large ethnic or immigrant groups do not belong to listed groups, and are not used to interacting with government agencies. They may even be fearful of so doing. A social profile, drawn up by an applied sociologist would give a clear insight into the community from which the stakeholders are being sought. A community assessment process such as that used by the California Department of Toxic Substances Control (Marxen, J. 2001. 6-4 to 6-15) would also be an effective tool here.

Public Participation Techniques

In the Republic of Ireland, legislation has generally required the proponent of a project, or applicant for a license to notify the public of her/his intentions in this regard. The regulatory authority then takes responsibility for conducting the consultative process. However, although at present the proponents are only 'encouraged' (for example in the EPA guidelines for EIA), to involve the public at the earliest stage of project design, the Aarhus convention (Sections 4-6) 'requires' it.

So, for the future, a whole range of public and private bodies will be involved in public participation, and in an increasing variety of circumstances. The following list of good practice principles (European Commission, 2000. p.239) should be considered before selection of specific methods.

General Guiding Principles

- o Plan early for public participation to establish the resources needed to support the process, and identify the limitations of finances.
- o Identify stakeholders and their legitimacy and/or representativenesss.
- o Identify the appropriate techniques to be used for each stage of the process, and provide information in a form that is comprehensible to the recipients.
- o Hold events at a time and venue to suit all participants.
- o Allow sufficient time for stakeholders to assimilate and respond to information.
- o Ensure that the inputs of stakeholders are integrated into any decisions made.

o Ensure feedback on all issues raised.

Which methodology to use depends on many factors, but the following questions adapted from (Cuff, J. 2001. p.18) (Acland, A. 2002. 3-2) will help in deciding.

o How complex is the issue to be discussed?

o What processes are already in place for resolving the issue and how well are they working?

o What processes have been tried previously and how well did they work?

o What is the geographical scale of the issue?

o At what levels of societal structure is the process to operate?

o What resources are available in time, money and in-house skills?

o Will external professional facilitators or consultants need to be employed?

With regard to this last question, apart from the costs, the employment of consultants can raise other issues. Consultants are only as good as their brief, and whilst the process may work well, (Cuff, J. 2001. p.131) the end product may be less of a success. The use of professional facilitators combined with stakeholder training in facilitation skills, early in the process, would enable a wider range of techniques to be used. It would also make the use of group-work, one of the most useful tools in concensus-seeking, more viable. (Cuff, J. 2001. p.49).

Good Practice Principles Specifically For Project Proponents (Acland, A. 2002, 3.2) include:

- As soon as possible focus on specifics, away from slogans.
- Acknowledge differences-then keep an open mind, thinking creatively to build on common ground.

- Uncertainty leads to distrust to fear to hostility, so clarify any uncertainty as early as possible. Be as open as possible and explain why, when you cannot be.
- Where legally possible, acknowledge past mistakes.
- Resolve internal issues prior to public participation.
- Address all dimensions of a situation; differences of needs and interests, behavior, and beliefs and values.
- Only make commitments you will be able to keep.
- Being open, consistent and honest in stakeholder relations is essential to building of trust.
- Flexibility in the choice of methods and use of same enables the creativity of the process.

Most important of all, it should be remembered that, stakeholder dialogue is often messy, disjointed and even chaotic at times. This is just how it should be-reflecting the real world in which we all have to operate (Acland, A. 2002, 3-2). So events should be made enjoyable. Having regular breaks for stretching, food and refreshments.

Process Planning

Many of the questions posed above can be answered during a process planning session (Wates, N. 2000. p.106) which allows stakeholders to determine the most suitable public participation methods for their particular situation.In order to do this, it is necessary for the participation coordinators to make an early attempt at identifying key stakeholders.

Record Keeping and Reporting

In order to maintain continuity and for confidence building, records of all sessions/meetings are kept and reported to all stakeholders, together with outcomes from the process. This is essential.

Public Participation Tools

What follows then is a survey of some of the tools most widely used in public participation. They are arranged in groups based on their general purpose and according to the four levels of participation outlined by Leach, S., and Wingfield, M. (1997, p.27).

Level 1 Education and information provision.

Level 2 Information and feedback.

Level 3 Involvement and consultation.

Level 4 Extended involvement.

These four levels categorise methods, according to the amount of power transfer from the responsible authority to the public. In level one, the public is passive, receiving information. In level two, feedback is requested and considered as part of the EDM process. Level three involves information, sharing and consultation, on all relevant aspects, before the final decision is made by the authority. At the fourth level, the authority can delegate power to a group of individuals to make decisions on its behalf. Clearly, due to legislative restrictions, level four processes will be rare, except in a very local context.

By contrast levels two and three will be widely used in the future. These levels are not, however, prescriptive, and some of the tools, such as interviews, can be combined with others at different levels. For a more detailed breakdown of the effectiveness of the different methods, see Canter, L.W. (1996).

The public participation tools described here are an amalgam of the author's experiences, and information from the following sources: (Canter, L.W. (1996), Gray, A.J. et al (2002) Yosie, T.F. and Herbst, T.D. (1998), IAP2(6) (2003), Ortolano, L. (1997), Connor, D.M. (2003), I.P.M.P. (2003) and Wates, N. (2000).

Level 1. Education and Information Provision

Printed Materials-Fact sheets, Newsletters, Brochures, Issue papers. The guiding rule here is KISS-Keep It Short and Simple, and embellish with good visual impact. The contents should recognise the pros and cons of an issue and explain the public's potential role in, and affect on, the project. An easy response method such as a postage paid card or free-phone hotline number will enable identification of stakeholders interested in involvement.

These materials can be targeted or general, and may reach a large audience. They may not, however, be read, and are limited in what information they can convey. Connor, D.M. (2003) estimates only 10% of the necessary information can be conveyed in this format. It is also difficulty to know if you have reached your audience, unless responses are generated.

Advertisements

Whilst advertisements in newspapers or magazines can potentially reach wide audiences, they are generally expensive and limited in the information they provide. It is essential to choose the pages, days and publications very carefully. Different publics read different publications! It is, as with other printed material, difficult to know if it has been read. Some feedback mechanism should be included.

Press Releases

The potential for reaching the public is great if the issue is of sufficient interest to the media, and can include requests for feedback. There may be problems of misrepresentation and lack of editorial control. The use of publicity stunts may be of use in drawing media attention to the less exciting areas of EDM. This is a cheaper option than advertising, though local newspapers will often do features on regular advertisers more readily.

Public Displays

Exhibits in public places have the advantage of enabling a wider public to view models and displays at their own choice of time and at their leisure. They are cheap in public buildings, but 42 to be more effective require the back up of personnel to explain details and/or provide literature.

They may also need to be duplicated or travel around the stakeholder constituency. Newspaper Inserts. To be successful an insert needs to be: of the best graphic design, in order to catch the reader's attention; in the publication on days when others are not; and preferably in the relevant local paper, where it is likely to have greater impact.

Newsletters

Although everyone will not read a newsletter it is an ongoing source of information with enough print space for some detail. It can also be used as an insert in the local paper. A newsletter also has great potential for feedback, via reply paid response forms, with for example prize draws attached.

Bill Stuffers

Any of the printed formats above could be included with utility bills, rates bills etc. Feedback possibilities can be included, these to be returned along with the bill payment.

Information Repositories

Usually, the regulatory authority makes EDM documents available at its offices and during office hours. This excludes, or makes difficult, access for many stakeholders. Libraries, as an alternative, are locations which have hours designed for public accessibility and whose 'raison d'etre' is in part the making available of information (EC1, 2000). Information repositories are, however, not widely made use of by the public. (IAP2 (6), 2003).

Site Visits

Small groups of stakeholders can learn more from first hand experiences than from any other means. It may, however, be difficult to identify a suitable site, that replicates all the issues faced by those in the EDM process.

Video

A production controlled by the proponent or the regulator may be seen as propaganda, but if produced with stakeholder involvement may be a good substitute for site visits, and can be shown widely over the stakeholder constituency. Whilst it is expensive to produce, a professional and credible video, once made, is not too expensive to reproduce, and so can be distributed to community groups, NGO's, employees etc.

Independent Technical Documents and Expertise.

Such documents could give credibility to a process, though they may be incomprehensible to the general public unless a non-technical summary is given. Public access to experts builds credibility and may help resolve disputes over disputed 'facts'. The reality exists, however that these technical experts may contradict information provided by the project proponent.

Level 2. Information Feedback

Public Meetings

This widely used method is described by (Connor, D.M. 2003) as, "the last of the blood-sports, and like the others, should be banned". Public meetings usually involve a speaker or panel giving a presentation, followed by a question and answer session. Sometimes these are broken down into smaller group discussions returning to plenary sessions. Gray, A.J. et al, (2002) suggest that public meetings offer an opportunity for stakeholders to meet, as well as demonstrating the willingness of the proponent to meet the other interested parties. However, they go on to say that they are one of the most complex and unpredictable methods. They can also be intimidating and can be hijacked by vocal individuals or interest groups. Badly handled, they can entrench opposition to a project, and division within a community. As a means of sharing information with those able and interested to attend, it is reasonably cost effective if run by well-trained staff.

Public Hearings

These formal occasions have a hearing officer, and a stenographer. Formal presentations from the public are received. This rather rigid format should according to (IAPZ (b) 2003) be avoided if possible, as it does not foster constructive dialogue and can perpetuate an 'us versus them' feeling, with little dialogue between participants. This format enables the public to speak without rebuttal, which is positive or negative depending on perspective.

The Internet

The use of the internet as a means of making available large amount of regularly up-dated reference material has been seen by many as the way forward in information provisions. The development of the European Commissions 'single access point', 'Your Voice in Europe' (Europa, 2003) is a good example. This portal site provides access to: every information source in the EU; consultations, where opinions are solicited on a wide range of policy-making and rule-making decisions; discussions with the EU leaders on-line; scrutiny of the working of the EU, with the possibility of giving anonymous feed back; and useful links to other sites and ways the public can be heard. There are, however, pitfalls in relying too heavily on this low cost innovative tool, and its potentially global audience. (Hacklay, M. 2002) found that only 10% of the total population of the U.K. had daily access to the Internet, with only 3% of those coming from the lowest income group, and 48% from the uppermost group. Hacklay also found, that only a small minority of the population, having access to the Internet, regularly entered the environmental information sites. This minority came mainly from the educated middle classes. He also concluded that the traditional media were the most used sources of information on the environment.

The Central Statistics Office, 'Fourth Quarter 2000 Survey' (C.S.O., 2001) showed 417,000 households in Ireland owned computers, with 262,700 having internet connections. This latter figure representing 20.4% of all households was the quadruple of the 1998 figure, and is likely to be much lower than that existing in 2003. Based on these figures, and Hacklay's findings, it is clear that caution should be used in relying on the internet as a means of communication.

Free-phone Telephone Lines

Well advertised, these phone lines can be used for the public to obtain information, ask questions or make comments about proposals or issues. The person on the line must be dedicated, well trained and have access to the information themselves, with a guaranteed call back time for complex answers that require research. This shows transparency and allows for up to date information to be easily available. Good examples of this type of service are: the Europe Direct, Free Phone 00800 678910; and the ENFO information line, Lo-call, 1890 200191.

Interviews

One to one meetings with stakeholders provide for a two-way flow of in-depth information and can be used to evaluate potential members of community advisory/liaison groups. Interviews are very useful for developing and refining public involvement and consensus building programmes. However, this can be a very time consuming process, if multiple interviews are scheduled. Closed meetings can also be treated with suspicion, especially by those stakeholders, which do not trust the independence of the process.

Surveys

Face to face surveys, conducted in a 'random' manner, can gather information from people who would not, or could not, become involved in other ways. As with all data gathering it is essential to have a clear purpose in asking the questions, and a use for the results achieved. This is an expensive method, but confidential surveys can elicit more candid results.

Response Sheets

As mentioned earlier, leaflets etc. can have reply paid cards attached, and these can be used to gather information on public concerns and preferences. These would generally have options for being included on a mailing list. There is, however, no statistical basis to the results, as the method is only as good as the distribution, presentation and in some cases the original mailing list. Results can be easily skewed.

Random Postal/Telephone Surveys

Research of this nature needs professional planning in order to attain statistically useful results, and can be labour intensive and expensive. The response rate to telephone surveys is much higher than to mail shots, but they are also more expensive to operate. That said, these surveys would generally reach a much wider cross section of the public, including those unlikely to attend meetings. The statistically tested results produced would also be more persuasive with all the stakeholders. These surveys are generally used for attitudinal research.

Internet Surveys

Precision in setting up a site for information exchanges is essential, otherwise chat rooms can generate far more input than can be looked at. The response rate is generally high but the user profile is limited by the 'digital divide,' as mentioned earlier, and so access is limited. There is of course no geographic control over the reach of the poll, it can be easily skewed and is very labour intensive.

Deliberative Polling

To measure informal opinion, randomly selected members of the public are brought together to listen to expert opinions and their own views on an issue. They are then confidentially questioned about their opinions. The result provides a representation of the publics considered judgment. This can take 2-3 days, involves up to several hundred people, requires the use of trained facilitators, and as a result can be very costly. Television cameras can be used to document the proceedings if the issue is important enough, and a resulting documentary used to air the debate more widely.

Teleconferencing

This enables face to face questions and answer sessions. This is useful where the stakeholders find it difficult to travel. This can be an expensive process, but also an inclusive one. Access to teleconferencing systems, may however be limited.

Presentations to Community Groups

At their regular meetings, social and civic clubs often invite guest speakers. The key here is to keep it short and simple (KISS), use visual aides and any other props that enable a lively presentation. The audience here may not have attended other functions, and this enables an expansion of mailing lists. The presenter has control of the information, but feedback can also be elicited. As the guest, the presenter has no input into who is invited to attend, and it is possible that project stakeholders may not be present..The audience may, however, become stakeholders.

Care must be taken to ensure the presentation is not too technical for the audience. Once prepared, the presentation can be used many times in the same format, adapting as necessary. Expert Panels. Members of local and national media are invited to publicly interview an 'expert' panel, followed by public participation. Efforts should be made to involve the stakeholders in choosing a neutral moderator, and setting the ground rules for the meeting.

This format encourages media education and gives the opportunity to dispel scientific misinformation, alongside a balanced discussion of the key issues. It does, however, require substantial organization, and may even enhance public concerns by increasing the visibility of the issues in an uncontrolled debate. The experience of the author is that, like public meetings, these are not too productive for the end game of environmental protection.

Field Offices

Where the stakeholder constituency is over a wide geographic area, the use of shop fronts in various accessible and much frequented locations, on stalls in enclosed shopping precincts and at festivals etc. can provide for a broader public outreach.

The provision of well trained staff, brochures, video and internet access points will all add to the effectiveness. Such traveling units can be brought to schools, third level institutions etc. This is a relatively expensive method, but provides opportunities for positive media coverage as well as providing for responsive, ongoing and focused communication.

Informal Small Group Meetings

These would usually be in a person's home within the neighbourhood, and involve a polite and appreciative member of staff. A coffee morning gives the relaxed setting conducive to effective dialogue. Clearly the numerical outreach is small, but the meeting being held on the communities own 'turf' and at their invitation creates a new relationship of equals.

Level 3. Involvement and Consultation

Workshops

These small public meetings (9-24 people) are designed to involve the participants in an active way. Wates, N. (2000) shows a typical format. They are a powerful tool for in-depth discussion, and with a good facilitator; the views of all those present will be heard and considered. It is also a good format for decision-making.

Workshops can be used to promote involvement and discussion at larger presentations, as well, with the attendees splitting up into small discussion groups, each with a facilitator, and reporting back to the main body.

This prevents domination of proceedings, by individuals or organizations. Hostile participants, by contrast, may see it as 'divide and conquer'. Workshops humanize project team members, promote public ownership of the process, and can be targeted at particular stakeholders. Before workshops commence, it is essential to know what is to be done with the inputs received. Workshops can also be used to 'thrash out' issues raised in other formats.

Focus Groups

These are a quick means of gauging what is likely to be the public reaction to a proposal. Involving a group of 15-25 invited participants, properly facilitated, this method can give a detailed understanding of people's perspectives, values and concerns. Selection of the participants so as to involve a good cross section of stakeholders is critical. Such a forum could be a one-day event, repeated as necessary.

The participants' needs for information etc. would have to be serviced. Conflicting viewpoints can be resolved, and personal respect engendered, in this as in all other small groups. If well planned and facilitated, Focus Groups are relatively inexpensive to organize.

Open House

Like any open day, this is the opportunity for the proponent, or regulator, to give a broad and detailed account of the project. The public are invited to a venue, preferably a neutral one in their own community, any time during the opening hours of 2-9.00 p.m., to enable maximum attendance. The individual is greeted at the door and the layout of information is explained. The members of the public then visit a number of 'stations', each illuminating a particular point and viewable by 6-10 people at a time. Each station is staffed by a resource-person, who can answer questions, and explain the exhibit. Visitors are asked to fill out comment sheets.

To make this a worthwhile exercise requires good publicity and the preparation of staff and materials, as well as time spent on the day. This makes it an expensive method, and one which, without added attractions, will not get much media attention. Conversely, it enables small group/one to one communication. It also allows team members to share information, and when needed, to answer visitors questions. The open house builds credibility. Open houses can be operated on the internet. Such open houses are cheap, accessible at all hours and not geographically limited. However, the 'digital divide', and the openness of the process make it difficult to use the inputs in any authoritative way, except as a source of ideas.

The Delphi Method

Following mail-shots, directed or random, with reply-paid questionnaires, the responses are collated, and then copies of the collated responses are sent to all those who responded for further comments, and so on. This enables consensus building, without the necessity for meetings and can also be done using combinations of telephone and post, or, with the caveats previously mentioned, by e-mail. It is a relatively cheap way of assessing the major issues relating to a project. The same methodology can be used to achieve expert consensus over contentious facts.

Level 4. Extended Involvement

Citizen Juries

Essentially an informal inquiry, this method involves 15-20 ordinary members of the public selected to represent a cross section of the community. They are brought together to consider a particular issue and receive evidence from expert witnesses, whom they can cross-examine. At the end of the process they produce a report setting out the views of the jury, including differences of opinion. Although its findings are non-binding, and the jury has no legal standing, the commissioning body must follow recommendations or explain why. In this way, issues can be considered in depth and in a credible way by, 'ordinary citizens'. Flaws can be pinpointed and public reaction gauged, and in a fairly short period of time. It is essential to plan the jury well, being clear, as to the resources necessary, and about what will be done with the report and its recommendations. This is a resource intensive approach, requiring the payment of jurors, a facilitator, expert's fees etc.

Advisory Groups

These can be statutory as in the case of the EPA (Section 27 of the EPA Acts, 1992) or not, and may be granted powers ranging from making recommendations to exercising leverage over final decisions. They are usually a small representative group of stakeholders who meet to discuss issues within a particular remit. Their existence is long term and not issue specific, and requires a large commitment from the participants and more resources than some of the other methods. Rules and responsibilities must be clearly defined in advance, and a consistently credible, transparent process, with third party facilitators, is essential. Potential committee members must be interviewed before appointment and the necessary support structures put in place. Participants gain a good understanding of the issues, processes and other viewpoints leading towards compromise. There is, however, the risk that the public will see the committee as too 'cosy' with the sponsor. The sponsor must be willing to take the groups advice seriously, and accept when consensus cannot be reached.

Task Forces

Generally issue-specific, a task force, made up of independent experts and stakeholders can identify, evaluate and solve particular problems. Such a group has great credibility and provides constructive opportunities for compromise. Time and labour expensive, task forces may not come to clear decisions or reach concensus.

Planning for Real

This innovative method was designed originally for communities to use in urban land use planning, but has been adopted for other situations. A good example of this latter was 'Planning for Flood Plains' (Cuff, J. 2001. p. 2.1.2.) a river basin management project. The process contained in the registered brand name 'Planning for Real' focuses on the construction of models and the use or priority cards so reducing the need for good verbal or written sills. This empowering process enables the community to identify their own aspirations and issues, bearing in mind the constraints of the real world (Wates, N. 2000. p.100). Dedicated staff time is needed and the complex 3.D models are expensive. Community Visioning. Most suited to developing general plans for a site area or organization, this has been much used by UK Local Authorities for Agenda 21 processes. This is a pre-planning process designed to collectively create imaginative shared visions for what the future could be in a community. Because there is no control over the outcome, the project proponent must be clear that considerable power has been handed to the community.

That said, this process is likely to lead to strong commitment to the outcome from all participants. Visioning can also be done via the internet, but there is no control over who makes the inputs, and restricts the process to internet users. A variation on the process is the 'Future Search Conference' used to focus on the future of an organization, a network of people, (IAP2 (b) 2003) or a community. As with most of these techniques, it is essential to use a trained facilitator. This method can involve hundreds of people simultaneously and each one is an 'expert'. A 2-3 day meeting, this may prove logistically challenging. It may also prove difficult to gain complete commitment from all stakeholders.

Evaluating the Process

Evaluation has been defined as 'a process of assessment which identifies and analyses the nature and impact of processes and programmes' (Interact, 2001). The evaluation of public participation is, however, a relatively rare activity (Gray, A.J. 2002. p.46). Yet it is an essential feedback mechanism during the process for the stakeholders, and post process for the funders, proponents, and stakeholders, to see, if the process is going well, or whether it was worth the effort.

As a learning process for future processes, clearly, record keeping and assessment are essential. Clearly then, it is not just the end result that is of importance in this assessment, but the journey to reach it. Most of the international evaluations use 'fairness; and 'competency' as underlying criteria (Webler; T. 1995), and most rely on 'generic' expert criteria. But Barnes, M. (1998) suggest that the evaluation process itself has to adopt participatory techniques if the criteria for assessment are to include those considered important by all the stakeholders.

Looking at the assessment of participatory processes for the UK Environment Agency, Petts. J. (2000) uses some 12 effectiveness criteria.

- o Clarity of objectives-is the connection made between purpose process and outcome?
- o Clarity of legal process-including limitations of outcome.
- o Consensus on Agenda and Procedures including choice of facilitator or chair person.
- o Input into assessment process.
- o Representativeness.
- o Inclusivity-are all barriers to participation minimized, and are non-participant stakeholders kept informed?
- o Deliberation-is dialogue open, inclusive, informal, detailed enough, educational and constructive?
- o Capability-are sufficient resources in time, money and staff training available?
- o Sound learning-is the process enjoyable and informative? Does it lead to increased consensus?
- o Actual Decision Responsiveness-is the process outcome included in the final decision made by the relevant authority?
- o Trust Enhancement-has the process increased trust between proponent and stakeholder for future participatory processes?

Perhaps a more useful tool, however, is that developed for the EU Life-Environment Project, 'Wise use of Flood Plains' (Cuff,J.,2001,pps 75-81). This could be used for an individual public participation method or for a whole public participation process. Figure 2. Cuff's Circle. Summary framework for evaluation of participatory processes. (Cuff, J., 2001) The categories are adapted here, where necessary, for a more general use by the author. The circular process, with 'outcomes' as its centre, and about which all else revolves, is a useful visual to maintain focus throughout the process. Objectives-These could be the objectives of the proponent and/or the objectives created as part of planning for the process. The two sets of objectives could be compared to see if participation objectives complement strategic ones. Did the objectives change over time and were they realistic. To what extent were objectives met. Contexts-Is the process connected to: for example rule making, long term planning, licensing, land use planning or a combination of two or more, e.g. a planning application with an EIS and an IPPC Licence process all on the one project? If so, how did the different processes interlink?

Other factors could be political (e.g. local elections) or geographical (e.g. a long powerline where jobs depend on it at one end, but people along its route resent it). Increased pressures on a process can be due to a need to act, brought about by external factors, e.g. flooding of the Somerset Levels (Cuff, J. 2001. p.77). Levels of Involvement-Using Arnstein's ladder, methodology can be assessed and explanations given for the level of power transfer used, in the light of the submissions made.

Inclusivity and Stakeholder Identification-This should also be measured in the light of the objections. Who was left out, or included? Were there complaints about this? Was a special effort made to include socially excluded groups? Were events timed to suit stakeholders? Was there a reluctance to participate? if so, why? How many, and what type of people actually took part?

A format. that could be used to evaluate individual participatory methods or processes. Scoring: 0= went badly, did not meet objectives, 10= all objectives met, total success. Participatory Methods/Processes-This assessment should be for each method used, and an objective set for each at its onset. The use of an H Diagram gives a good picture of the pros and cons of an event. All aspects of the methodology should be scored from 1-10 to show strengths and weaknesses, and recommendations for the future use of same. Innovation-Any new methods could be shown and assessed here, for possible use in the future.

Commitment to using Participation Process Results-Was it clear at the outset what limitations the participation process had in influencing the decision-making? What actual use was made of the process outcomes? Were participants informed of the use of their inputs? If so, when and how? Inputs-A break down of the costs of the process into individual events, and then into their constituent costs, is needed to assess the value-for-money represented by the different methods. Was lack of finance an obstacle? Was the use of consultants value for money? Outputs-This is a measure of the number and quality of outputs and their effectiveness, e.g. posters, exhibitions, questionnaires, etc. Outcomes-These stem mainly from 5 and 9 (figure 2, above) and are measured against the baseline of the pre-process situation. Firstly, there are the strategic outcomes. Has a decision/ recommendation been made, and is it based on a consensus? Secondly, there are the cultural outcomes. For stakeholders these include:

- o Change in the attitude to participatory processes.
- o Ownership of the process o Increased trust in the proponent
- o Changes in values or attitudes

For organizations:

- o Are they more transparent?
- o Are they more at ease with public participation?
- o Do they see the process as worthwhile?

There will be other outcomes in the many circumstances in which this assessment process could be used.Wates, N. (2000. p.100) describes how to organize and run an effective 'review session' involving all stakeholders.

Conflict Resolution

Having pursued a participatory programme what can be done if the stakeholders cannot reach concensus? The use of strategies known as ADR (Alternative Dispute Resolution) and based on 'joint gains' problem solving, mediations, facilitation and consensus building offer promise for many cases (Adler, P.S. et al., 2003). Bingham, G. (1986) showed that 78% of the 161 environmental disputes studies, which used mediation-based ADR, ended in agreement amongst the parties. It is, however, not a panacea, and each case should be judged on its merits.

The US EPA uses the definition of ADR in the Administration Dispute Resolutions Act of 1996 "any procedure that is used to resolve issues in controversy, including but not limited to conciliation, facilitation, mediation, fact finding, minitrials, arbitration and the use of ombuds or any combination thereof. All the methods listed above involve a neutral third party who assists others in designing and conducting a process aimed at reaching agreement. All aspects of ADR are voluntary, including the decision to participate, the type of process used, and the content of any final agreement. The ADR process and practices are beyond the scope of this study and generally would only be undertaken using specialist EDM mediators or facilitators, however, it is worth looking at the underlying principle. The conditions, under which ADR is useful, and most likely to be successful are listed by Otolano (1997. p.417) as:

- A high uncertainity in timing and outcome. Stakeholders who perceive other quicker or more certain routes to achieve their objectives all have little motivation to become involved in ADR.
- Equivalent distributions of power. Here power is described as the ability to prohibit an action desired by one or more adversaries.
- Mediation is more promising than BATNA (Best Alternative to a Negotiated Agreement). As long as the prospective outcome from negotiation is more attractive than BATNA, a stakeholder has an incentive

to continue negotiating. If the prospects for success for negotiation drop than the BATNA might be the courts, lobbying or direct action.

- o Disputing parties are at an impasse.
- o Litigation is ineffective.

Ortalano, L. (1997, p.411) goes on to describe the ADR techniques known as 'principled negotiations'. This is characterized by four main points, which are a useful insight into problem solving during any EDM process.

- o Separate the people from the problem
- o Focus on interests not positions
- o Invent options for mutual gains
- o Use objective criteria for evaluating the agreement.

Implicit in these is the recognition of the four main types of conflict delineated by Creighton, J.L. (1981. pp 454-455).

- o Cognitive conflict-differences over the facts.
- o Values conflicts-a dispute over goals, for example a dispute over whether a particular outcome is desirable (undesirable) or should occur or not.
- o Interest conflicts-Cost and benefits are rarely evenly distributed so that some may have a greater interest in a particular outcome. It is possible then to have agreement on facts and on values but still disagree on interest.
- o Relationship conflict-Communication involves both facts and relationships (how much a person is valued, accepted etc.). The EDM process itself may favour those with power and money and so communicate a lesser value to other stakeholders, leading to conflict on emotionally motivated grounds other than facts, values or interests.

3

ECONOMIES OF NATURAL RESOURCES

INTRODUCTION

The dominant ideology of development. which guides development activities almost exclusively, has been classically concerned only with the use of natural resources for commodity production and capital accumulation. It ignores the resource processes that have been regenerating natural resources outside the realm of human existence. It also ignores the vast resource requirements of the large number of people whose needs are not being satisfied through the market mechanisms. The ignorance or neglect of these two vital economies of natural resources, the economy of natural processes and the survival economy, explains why ecological destruction and threat to human survival have remained hidden negative externalities of the development process. To make good for this shortcoming it is necessary to comprehend the place of natural resources in all the three economies.

A new and holistic relationship between economics and ecology has to depend on a holistic understanding of the natural resource process and utilizations associated with human societies and the natural ecosystems. The incompetence of modern economics in dealing with natural resources in their ecological totality has been voiced by many. The most penetrating description, however, comes from Georgescu Roogen who wrote: The no deposit no return analogy benefits the businessman's view of economic life. For, if one looks only at money, all one can see is that money just passes from one hand to another: except by regrettable accident it never gets out of the economic process. Perhaps the absence of any difficulty in securing raw materials by those countries where modern economics grew and flourished was yet another reason for economists to remain blind to this crucial economic factor. Not even the wars the same nations fought for the control of the world's natural resources awoke the economists from their slumberer.

While trade and exchange of goods and services have always characterized human societies, the elevation of the market to the position of the highest organising principle of society led to the neglect of the other two vital economies in development thought. The hidden negative externalities of the development processes governed by the principles of the market have, thus, created new forms of poverty and underdevelopment. Various case studies described in this volume will substantiate such a claim. The major problem is that when exclusive attention is being given to monetary flows, requirements of natural resources not backed up by suitable purchasing power cannot be registered on the economic scene. As a result, specially in the context of Third World countries, the place of natural resources in the economy of natural resource production (or nature's economy) or in the survival economy of non-market consumption for the biological sustenance of the marginalised poor gets completely ignored.

The political economy of ecology movements cannot be understood without a clear comprehension of the place of natural resources in the three distinct economies. Ecology movements are the first indicators of compatibility and conflict among the three competing demands on natural resources.

In this way the articulation of these three economies provides the foundation of a framework for an ecologically sustainable and equitable process of economic development that ensures survival and does not threaten it. The benefits and costs associated with development projects thus need to be evaluated not only in terms of the framework of the market economy but also in terms of the other two economies associated with natural resources.

THE ECONOMY OF NATURAL ECOLOGICAL PROCESSES

The terms ecology and economy are rooted in the same Greek word 'oikos' or household. Yet in the context of market-oriented development they have been rendered contradictory: 'Ecological destruction is an obvious cost for economic development'-a statement which is often repeated to ecology movements. Natural resources are produced and reproduced through a complex network of ecological processes. Production is an integral part of this economy of natural ecological processes but the concepts of production and productivity in the context of development economics have been exclusively identified with the industrial production system for the market economy. Organic productivity in forestry or agriculture has also been viewed narrowly through the production of marketable products of the total productive process.

This has resulted in vast areas of resource productivity, like the production of humus by forests, or regeneration of water resources, natural evolution of genetic products, erosional production of soil fertility from parent rocks, remaining beyond the scope of economics. Many of these productive processes are dependent on a number of ecological processes. These processes are not known fully even within the natural science disciplines and economists have to make tremendous efforts to internalize them. Paradoxically, through the resource ignorant intervention of economic development at its present scale, the whole natural resource system of our planet is under threat of a serious loss of productivity in the economy of natural processes.

At present ecology movements are the sole voice to stress the economic value of these natural processes. The market-oriented development process can destroy the economy of natural processes by over exploitation of resources or by the destruction of ecological processes that are not comprehended by economic development. And these impacts are not necessarily manifested within the period of the development projects. The positive contribution of economic growth from such development may prove totally inadequate to balance the invisible or delayed negative externalities stemming from damage to the economy of natural ecological processes. In the larger context, economic growth can thus, itself become the source of underdevelopment. The ecological destruction associated with uncontrolled exploitation of natural resources for commercial gains is a symptom of the conflict between the ways of generating material wealth in the economies of-market and the natural processes. In the words of Commoner: 'Human beings have broken out of the circle of life driven not by biological needs, but the social organisation which they have devised to 'conquer' nature: means of gaining wealth which conflict with those which govern nature."

THE SURVIVAL ECONOMY

Modern economics and the concept of development cover a miniscule portion in the history of economic production by human beings. The survival economy has given human societies the material basis of survival by deriving livelihoods directly from nature through self-provisioning mechanisms. In most Third World countries large numbers of people are deriving their sustenance in the survival economy in ways that remain invisible to market oriented development. Within the context of a limited resource base the destruction of the survival economy takes place through the diversion of natural resources from directly sustaining human existence to generating growth in the market economy. Sustenance and basic needs satisfaction is the organising principle for natural resource use in the survival economy whereas profits and capital accumulation are the organising principles for the exploitation of resources for the market. Human survival in India even today is largely dependent on the direct utilisation of common natural resources."

Ecology movements are voicing their opposition to the destruction of these vital commons so essential for human survival. Without clean water, fertile soils, and crop and plant genetic diversity economic development will become impossible. Sometimes by omission and sometimes by commission formal economic development activities have impaired the productivity of common natural resources which has enhanced the contradiction between the economy of natural processes and the survival economy.

The organising principles of economic development based on economic growth renders valueless all resources and resource processes that are not priced in the market and are not inputs to commodity production. This premise very often generates economic development programmes that divert or destroy the resource base for survival.

While the diversion of resources, like diversion of land from multipurpose community forests to monoculture plantations of industrial tree species, or the destruction of common resources, or the diversion of water from staple food crops and drinking water needs to cash crops are frequently proposed as programmes for economic development in the context of the market economy, they create economic underdevelopment in the economies of nature and survival. Ecology movements are aimed at opposing these threats to survival from market based economic development. Thus in the Third World, ecology movements are not the luxury of the rich; they are a survival imperative for the majority of people whose survival is not taken care of by the market economy but is threatened by its expansion.

The political foundation of ecology movements lies in their capacity to enlarge the spatial, temporal and social bases for the evaluation of economic development projects- in their capacity to bring into the picture all the three economies described earlier. A new economics of development will emerge only when these three economies can be conceptualized within a single framework.

TECHNOLOGY CHOICE TOWARDS HOLISTIC ECOLOGICAL CRITERIA

When economic development programmes are viewed from the perspective of all the three economies, a clearer view of the political economy of conflicts over natural resources is expected to emerge. In the dominant mode of economic development, perceived within the framework of the market economy, mediation of technology is assumed to lead to the control of larger and larger quantities of natural resources, thus turning scarcity into abundance and poverty into affluence: Technology, accordingly is viewed as the motive force for development and the vital instrument that guarantees freedom from dependence on nature' The affluence of the industrialized west is assumed to be associated exclusively with this capacity of modern technology to generate wealth.

The concept of technology per se as a source of abundance and freedom from nature's ecological limits are based in part on the limitations of the market economy in understanding in a holistic manner, the same resources which it exploits. Only when development processes are viewed in the holistic perspective of all the three economies can the scarcities and underdevelopment associated with abundance and development be clearly seen. Most resource-intensive technologies operate in the enclaves with enormous amounts of various resources coming from diverse ecosystems which are normally far away. This long, indirect and spatially distributed process of resource transfer made possible by energy-intensive long distance transportation, leaves invisible the real material demands of the technological processes of development.

The spatial separation of resource exhaustion and the creation of products have also considerably shielded the inequality creating tendencies of modern technologies. Further, it is simply assumed that the benefits of economic development based on these modern technologies will automatically percolate to the poor and the needy and growth will ultimately take care of the problems of distributive justice. This would, of course, be the case, if growth and surplus were in a sense absolute and purchasing power existed in all socio-economic groups. None, however, is correct. Surplus is often generated at the cost of the ecological productivity of natural resources or at the cost of exhausting the capital of non-renewable resources. For the poor, the only impact of such economic activity often is the loss of their resource base for survival.

It is thus no accident that modern, efficient and 'productive' technologies 'creased within the context of growth in market economic terms are associated with heavy social and ecological costs. The resource and energy intensity of the production processes they give rise to demands ever increasing resource withdrawals from the natural ecosystems. These excessive withdrawals in the course of time disrupt essential ecological processes and result in the conversion of renewable resources into non-renewable ones. Over time, a forest provides inexhaustible supplies of water and biomass including wood, if its capital stock, diversity and hydrological stability are maintained and it. is harvested on a sustained yield basis.

The heavy and uncontrolled market demand for industrial and commercial wood, however, requires continuous over-felling of trees which destroys the regenerative capacity of the forest ecosystems and over time converts these forests into non-renewable resources. Sometimes the damage to nature's intrinsic regenerative capacity is impaired not directly by over-exploitation of a particular resource but indirectly by damage caused to other natural resources related through ecological processes. Thus under tropical monsoon conditions, over-felling of trees in catchment areas of streams and rivers not only destroys forest resources, but also stable, renewable sources of water. Resource-intensive industries do not merely disrupt essential ecological processes by their excessive demands for raw materials; they also destroy and disrupt vital ecological processes by polluting essential resources like air and water. In the words of Rothman: 'the private economic rationality of the profit seeking business enterprise is a murderous providence because it cannot guarantee the optimum use of resources for society as a whole. It cannot avoid continually creating situations which cause the pollution of an environment

In the context of resource scarcity where most resources are already being utilised for the satisfaction of survival needs, further diversion of resources to new uses is likely to threaten survival and generate conflicts between the demands of economic growth and the requirements of survival. It, therefore, becomes essential to evaluate the role of new technologies in economic development on the basis of their resource demands and conflict with the demands of survival. The productivity of 8 technology in the perspective of human survival must distinguish outputs in terms of their potential for satisfaction of vital or non-vital needs, because on the continued satisfaction of vital needs depends human survival. As Georgescu-Roegen points out.

There can be no doubt about it. Any use of the natural resources for the satisfaction of non-vital needs means a smaller quantity of life in the future. If we understand well the problem, the best use of our iron resources is to produce plows or harrows as they are needed, not Rolls Royces, not even agricultural tractors.

In the context of the market economy, the indicators of technological efficiency and productivity are totally independent of the difference between the satisfaction of basic needs and luxury requirements. between resources extracted by ecologically sensitive or insensitive technologies or of the nature of the contribution of economic growth to diverse socio-economic categories. In the context of a highly non-uniform distribution of purchasing power and scanty knowledge of or respect for ecological processes, economic growth depends on production and consumption of nonvital products. The expansion of the formal sector of the economy for the production of non-vital goods often leads to further diversion of vital natural resources. For example, water-intensive production of flowers or fruits for the lucrative export market often results in water scarcity in low rainfall areas. In a world with a limited and shrinking resource base, and in the economic framework of a market economy, non-vital luxury needs are fulfilled at the cost of vital survival needs. The high powered pull of the purchasing capacity of the rich of the world can draw out necessary resources in spite of resource scarcity and resulting conflicts.

This complete lack of recognition of the resource needs of the survival economy nature's economy in the current paradigm of development economics shrouds the political issues arising from resource transfer and ecological destruction. For the economic sector based on 'efficient modern technologies', this provides an ideological weapon for increased control of the sponsors of economic development over the total natural resource endowments of the countries concerned.

The ideological and limited concept of 'productivity' of technologies has been universalised with the consequence that all other costs of the economic process become invisible. The invisible forces which contribute to the increased 'productivity' of a modern farmer or factory worker emanate from the increased consumption of non-renewable natural resources. Lovins has described this as the amount of 'slave' labour at present at work in the world. According to him, each person on earth, on an average, possesses the equivalent of about fifty slaves, each working forty hours a week. Man's annual global energy conversion from all sources (wood, fossil fuel, hydroelectric power, nuclear) at present is approximately 8 x 10 (12) watts. This is more than twenty times the energy content of the food necessary to feed the present world population at the FAO standard per capita requirement of 3,600 cals per day.

In terms of workforce, therefore, the population of the earth is not 4 billion but about 200 billion, the important point being that about 98 per cent of them do not eat conventional food. The inequalities in the distribution of this 'slave' labour between different countries is enormous, the average inhabitant of the USA, for example, having 250 times as many 'slaves' as the 'average Nigerian'. And this, substantially is the reason for the difference in efficiency between the American and Nigerian economies: it is not due to the differences in the average 'efficiency' of the people themselves. There seems no way of discovering the relative efficiencies of Americans and Nigerians: If Americans were short of 249 of every 250'slaves' they possess, who can say how 'efficient' they would prove themselves to be.

The increase in the levels of resource consumption is taken universally as an indicator of economic development. If the present level of resource consumption in the USA is accepted as the development objectives of India, the total resource demands of 'developed' India can be calculated by multiplying the current resource consumption by a factor of 250. Neither our forests nor our fields or rivers can sustain such a 'development'. When per capita resource consumption is considered, the Malthusian argument relating population with resource scarcity does not hold good. More significant than the population factor is the total resource factor. Thus, although many countries of the South have a much larger population than those of the North, the industrialized of the world consumes more grain than all the other three-quarters put together. This high consumption is due to the fact that intensive livestock production in industrialized countries accounts for 67 per cent of their total grain consumption. This Efficient' process of livestock management for the production of meat, as reported by Odium requires 10 calories of energy input to produce one calorie of food energy.

The energy subsidy provided by the capital stock of the earth's non-renewable resources makes a resource inefficient process appear as efficient in the market economy. It is interesting to note that even in the West, nearly a century ago one calorie of food was produced by using a fraction of a calorie of energy input. The same is true in the economics of water resources use in modern agriculture. When the production of high yielding varieties of seeds is evaluated, not on productivity per unit land (tons/ha) but per unit volume of water input (tons/le lit), these miracle seeds of the Green Revolution are seen as two to three times less efficient in food production than, say, the millets. The results of evaluation of the technological efficiency of processes associated with economic development, when reexamined on a holistic basis and optimised against all resource inputs, would generally lead to the conclusion that: 'the much talked of efficiency of widely practiced high technology is not intrinsically true. They are, in fact, highly wasteful of materials and pollutive (that is, destructive to the productive potential of the environment)'.

New technologies in the market economy are innovated for profit maximization and not to encourage resource prudence per se. The extent of inefficiency in the utilisation of natural resources with production processes based on resource-intensive technologies, can be illustrated with the production of soda ash, an important industrial material. In the Solvay process for the production of soda ash. the two materials used are sodium chloride and limestone.

The entire limestone used in the process ends up as waste material, 25 per cent of the sodium chloride is lost as unreacted salt. From the balance 75-80 per cent, the acidic half is lost and only the basic half goes into the final product. Therefore only 40 per cent of the raw materials consumed are actually utilised. The waste products pollute land and water resources systems. The economy of the process is artificially made good by concessions in procuring limestone, salt and fuel and further concessions in respect of land, transport, etc. It is these subsidies for natural resources which make the counter-productive processes appear efficient.

Referring to the technology of production of frozen orange juice Schnaiberg made the following remarks:

What is true of the unobtrusive shift from fresh oranges to frozen orange juice is typical of most transitions from traditional to late industrial technologies. The majority of these become more energy intensive: the energy content of all the necessary production processes increases per unit produced.... The hall mark of modern technology is its typical labour saving quality-not its energy saving aspect."

Guided by a narrow and distorted concept of efficiency and supported by all types of subsidies, technological change in market economy-oriented development continues in the direction of resource intensity, labour displacement and ecological destruction. The long-term continuation of such processes will lead to the destruction of the resource base of the survival economy and to human labour being rendered dispensable in the production processes of the market economy. The partisan assumptions of modern economic development which cannot internalise the economy of natural processes and the survival economy are thus being raised to the level of universality. As a result, with the expansion of economic development in Third World countries, the resource-intensive and socially partial development is leading to social instability and conflicts. While ecology movements in the industrially advanced countries are directed against more recent threats to survival like pollution, ecology movements in Third World countries have a much longer history related to resource exhaustion and ecological degradation of natural ecosystems. It is in these countries that the holistic ecological criteria for technology choice is needed most urgently.

The process of transformation and utilisation of natural resources for the satisfaction of societal needs determines the economic organization of human societies. At various stages of development, the dominant patterns of utilisation of natural resources have been guided by the dominant pattern of scientific knowledge, and through the generation and use of technologies that actually bridge the gap between natural resources and human needs and requirements.

A special characteristic of human societies is that they can make deliberate choices between different ways of using resources and satisfying needs. The existence of plurality of alternatives in resource use for economic development creates the need for a selection criteria to make rational decisions about the use of natural resources and technological change. A dialectical relationship exists between the criteria of technology choice and the nature of science and technology developed in response to the criteria.

Traditional societies as well as modern scientific-industrial societies have adopted different systems of science and technology which differ primarily in the criteria of choice or rationality that guides resource use patterns for human needs satisfaction. The characterization of certain societies as primitive and unscientific is, thus, sociologically and epistemologically unfounded. The fact that values and rationality criteria of one form of social organization generate a particular type of science and technology matched to a particular criteria of scientificity does not imply that other social organisations lack a scientific basis for their economic activities.

If sustainable utilisation is the objective that guides the criteria of choice for a development strategy, a resource prudent technological path (T.) is rationally chosen. If maximization of the growth of man-made processes and increasing the productivity of labour is the objective, then a more resource-intensive path (T2) which is the integration of a large number of smaller technologies (t1) and in which increased resource and energy inputs allows the increase in labour productivity, is rationally chosen. In this process a large amount of secondary resources (R2-R6) are additionally required.

Traditional societies in all their diversity have, in general, shared a common set of characteristics. They have used natural resources prudently to satisfy minimum needs sustainably over centuries. Such resource use was based on.

Resource flow in resource prudent t1 and resource.-intensive t2 technology chains

1. A knowledge system with an ecological understanding of nature.
2. A technological system for processing resources to satisfy human needs with minimum resource waste.
3. Rationality criteria for demarcating vital and non-vital needs and between resource destructive and resource enhancing technologies.

Traditional world views and practices deterred over-exploitation of natural resources at all levels. As they were based on ecological perceptions of nature and guided by restraints in resource use, they used technologies which prevented ecological disruption. Modernisation of traditional societies in its present form has, by and large, been taken as synonymous with the substitution of indigenous science and technology systems by the modern western system. In this manner the resource-intensive western pattern of resource use is thrust on non-western societies through modernisation.

Modern western scientific knowledge, however, differs from indigenous knowledge systems in three important ways:

1. Modern western scientific knowledge is reductionist and fragmented.
2. Modern western technological systems are based on reductionist science and are generally more resource-intensive.
3. There are no criteria of rationality or technology choices to evaluate modern science and technology on the basis of resource use efficiency or need satisfaction capability.

These characteristics of modern western science and technology systems breaks the chain, beginning with natural resources and ending in the satisfaction of human needs and demands, into small fragments of individually identifiable economic activities. This provides justification for the resource intensity of the dominant paradigm of economic development and technological change, and thus leads to ecological instabilities.

Ecological crises are thus inevitable products of economic activities which are propelled towards longer and more complex and resource-intensive technological chains (T2) for the satisfaction of older needs (N.). Only individual segments(l) of the whole technological chain are examined from the narrow criteria of labour productivity. The situation is best exemplified in the case of food production. While indigenous and traditional food production practices used about half a calorie of energy to produce 1 calorie of food, the present mechanised and chemical farming techniques use 10 calories of energy to produce 1 calorie of food. These characteristics of contemporary scientific industrial development are the primary causes for the contemporary ecological crises. The combination of ecologically disruptive scientific and technological modes, and the absence of rationality criteria for evaluating scientific and technological systems in terms of resource use efficiency, has created conditions where society is increasingly propelled towards ecological instability and has no rational and organised response to arrest and curtail these destructive tendencies.

ECOLOGY AND THE POLITICS OF KNOWLEDGE

The paradigm of modern science has evolved in the last few centuries in an environment where all economic activities were aimed at maximising the productivity of man-made processes in individual sectors of the economy. This led to the development of modern technologies with highly negative externalities which remained invisible within the conceptual framework of modern science and economics. This shortcoming emanates from three basic fallacies of modern scientific knowledge:

1. It identifies development merely with sectoral growth, ignoring the underdevelopment introduced in related sectors through negative externalities and the related undermining of the productivity of the ecosystem.
2. It identifies economic value merely with exchange value of marketable resources. ignoring use values of more vital resources and ecological processes.
3. It identifies utilization merely with extraction. ignoring the productive and economic functions of conserved resources.

Development planning based on these false identifications tends to create severe ecological problems because of its inability to recognise ecosystem linkages and the ecological processes operative in the natural world. The ecological relationships between the sectors of natural resources contribute to essential ecological processes which are frequently found to be vital for human survival. Thus, the stability of ecological processes is not merely a matter of aesthetics. An incomplete understanding of the material and economic values of ecological processes leads to the destruction of the material conditions for economic development and eventually survival.

Since the availability of essential and vital resources for survival is dependent on the maintenance of essential ecological processes, economic activities which generate sectoral growth in the shortterm by destroying the essential ecological processes cannot lead to development in the long run. On the contrary, by decreasing the productivity and availability of vital resources, they initiate the process of underdevelopment.

When the natural world is viewed ecologically as a system of interrelated resources which maintain the material basis for human sustenance, economic values can no longor be perceived merely as exchange values in the market. Economic values in the ecological perspective are not always equivalent to their exchange value in the market, evaluated without any significance to their use value.

As a corollary, natural resources can have economic utility that cannot be quantified through the exchange value in the market. Such economic utility includes the maintenance of essential ecological processes that support human survival and, thus, all economic activities. The economic utilisation of resources through extraction may, under certain conditions, undermine and destroy vital ecological processes leading to heavy but hidden diseconomies. The nature of these diseconomies can be understood only through the understanding of ecological processes operating in nature.

The economics of sustenance and basic needs satisfaction is, therefore, linked with ecological perceptions of nature. The economics of sectoral growth on the other hand is related to reductionist science and resource wasteful technologies which are productive in the narrow context of sectoral and labour inputs, but may be counter-productive in the context of the overall economic base of natural resources.

The case studies in the following chapters are only representative of thousands of such cases seen everywhere. They reveal a certain pattern of contemporary economic development which can be identified thus:

1. Development has been equated only with the growth of manufacture in individual industrial sectors and with the increase in productivity of only man-made processes.
2. This sectoral growth of man-made processes has also led to ecological destruction of the natural resource base, affecting negatively other sectors of the economic system. This leads to the decay of systems productivity of all productive processes, man-made and natural.

As a result of this limitation of contemporary economics, economic development has, consequently, been taken to be synonymous with growth. The higher the rate of sectoral growth, the higher is the index of economic development. Possible ecological destruction caused by the resource intensity of sectoral growth that is guided purely by non-ecological economic considerations, has never been introduced in the processes of planning for economic development. The benefit-cost analysis of development projects has thus externalized those ecological changes and is incomplete in three important ways:

1. It deals with benefits and costs as profits and losses in financial ferms.
2. It deals with benefits and costs only in the narrow sectoral perspective and ignores costs generated by inter-sectoral linkages.
3. It deals with benefits that are largely available to more visible and economically powerful groups and ignores costs that are borne by the less visible and

economically weaker groups. These costs and the associated underdevelopment are thus made invisible in modern economic analysis.

The utilisation and management of natural resources in India has so far been guided by the narrow and sectoral concept of productivity and restricted benefit-cost analysis. This narrow concept of productivity and benefit-cost analysis has blocked the conceptualization of the criteria of rationality of technology choice which maximizes needs satisfaction while minimising resource use, thus maximising systems productivity.

For example, the clear felling of natural forests in the catchments of rivers, and planting of industrial species of trees has been justified on the grounds of increasing productivity of forests. This concept of productivity is, however, only related to productivity of industrial timber, while forests produce other forms of biomass, like fodder and green mulch, or maintain productivity of soil and water resources. The direct impact of the clear felling of catchment forests on agricultural production through its destructive impact on soil and destabilisation of the hydrological balance is not taken into account in the calculations of the benefits and costs associated with forests. Regular floods and droughts, which are the consequences of irrational land and water management, are branded as natural disasters for which the whole nation pays heavily. Consequently, the poor and marginal groups which depend on agriculture for their livelihood face increasing impoverishment and poverty. This thrusting of negative externalities on the poor and marginal groups directly leads to the polarisation of society into two groups. One group gains from the process of narrow sectoral growth, while the poor and marginalised majority suffer because of the ecological destruction of natural resources on which they depend for survival.

The dialectical contradiction between the role of natural resources in production processes to generate growth and profits and their role in natural processes to generate stability is made visible by movements based on the politics of ecology. These movements reveal that the perception, knowledge and value of natural resources vary for different interest groups in society. The politics of ecology is thus intimately linked with the politics of knowledge. For subsistence farmers and forest dwellers a forest has the basic economic function of soil and water conservation, energy and food supplies, etc. For industries the same forest has only the function of being a mine of raw materials. These conflicting uses of natural resources, based on their diverse functions, are dialectically related to conflicting perceptions and knowledge about natural resources. The knowledge of forestry developed by forest dwelling communities therefore evolves in response to the economic functions valued by them. In contrast, the knowledge of forestry developed by forest bureaucracies, which respond largely to industrial requirements, will be predominantly guided by the economic value of maximising raw material production.

The way nature is perceived is therefore related to the pattern of utilisation of resources. Modern scientific disciplines which provide the currently dominant perspectives of nature have generally been viewed es 'objective', 'neutral' and 'universally valid'. These disciplines are, however, particular responses to particular economic interests. This economic determination influences the content and structure of knowledge about natural resources which, in turn, reinforces particular forms of resource utilisation The economic and political values of resource use are thus built into the structure of natural science knowledge.

4

ENVIRONMENTAL ECONOMICS RELATIONSHIP TO OTHER FIELDS

INTRODUCTION

Environmental economics is related to ecological economics but there are differences. Most environmental economists have been trained as economists. They apply the tools of economics to address environmental problems, many of which are related to so-called market failures—circumstances wherein the "invisible hand" of economics is unreliable. Most ecological economists have been trained as ecologists, but have expanded the scope of their work to consider the impacts of humans and their economic activity on ecological systems and services, and vice-versa. This field takes as its premise that economics is a strict subfield of ecology. Ecological economics is sometimes described as taking a more pluralistic approach to environmental problems and focuses more explicitly on long-term environmental sustainability and issues of scale.

These two groups of specialists sometimes have conflicting views which can often be traced to the different philosophical underpinnings of the two fields. Some ecologists subscribe to deontological ethical systems; other economists subscribe to teleological ethical systems. Neither ethical system can be demonstrated to be right or wrong, but they may sometimes have different implications for environmental policy. Environmental economics is viewed as relatively more pragmatic in a price system; ecological economics as relatively more idealistic as it supposedly does not use money to arbiter decision making as much.

Another context in which externalities apply is when globalization permits one player in a market who is unconcerned with biodiversity to undercut prices of another who is - creating a "race to the bottom" in regulations and conservation. This in turn may cause loss of natural capital with consequent erosion, water purity problems, diseases, desertification, and other outcomes which are not efficient in an economic sense. This concern is related to the subfield of sustainable development and its political relation, the anti-globalization movement.

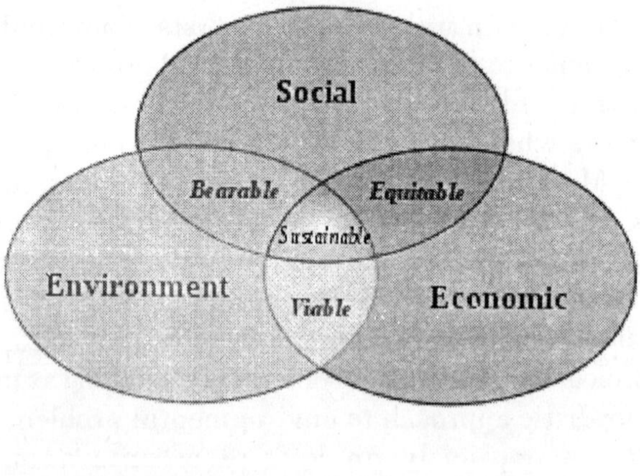

THE THREE PILLARS OF SUSTAINABILITY

Environmental economics was once distinct from resource economics. Natural resource economics as a subfield began when the main concern of researchers was the optimal commercial exploitation of natural resource stocks. But resource managers and policy-makers eventually began to pay attention to the broader importance of natural resources (e.g. values of fish and trees beyond just their commercial exploitation;, externalities associated with mining). It is now difficult to distinguish "environmental" and "natural resource" economics as separate fields as the two became associated with sustainability. Many of the more radical green economists split off to work on an alternate political economy.

Environmental economics was a major influence for the theories of natural capitalism and environmental finance, which could be said to be two sub-branches of environmental economics concerned with resource conservation in production, and the value of biodiversity to humans, respectively. The theory of natural capitalism (Hawken, Lovins, Lovins) goes further than traditional environmental economics by envisioning a world where natural services are considered on par with physical capital.

The more radical Green economists reject neoclassical economics in favour of a new political economy beyond capitalism or communism that gives a greater emphasis to the interaction of the human economy and the natural environment, acknowledging that "economy is three-fifths of ecology" - Mike Nickerson.

These more radical approaches would imply changes to money supply and likely also a bioregional democracy so that political, economic, and ecological "environmental limits" were all aligned, and not subject to the arbitrage normally possible under capitalism.

ECONOMIC DEVELOPMENT AND ENVIRONMENTAL CONFLICTS IN INDIA

The indigenous modes of natural resources utilisation were sensitive to the limits to which these resources could be used It is said that the codes of visiting important pilgrim centres Badrinath in the sensitive Himalayan ecosystem, included a maximum stay of one night so that the temple area would not put excess pressure on the local natural resources base. In the precolonial indigenous economic processes, the levels of utilisation of natural resources were not significant enough to result in drastic environmental problems. A characteristic of indian civilization has been its sensitivity to natural ecosystems. vital renewable natural resources like vegetation, soil and water were managed and utilised according to well defined social norms that respected the known ecological processes.

There were useful social norms for environmentally safe resource utilisation and people protested against the destructive use of resources even by kings. A major change in the utilisation of natural resources of India was introduced by the British who linked the resources of this country with the direct and large nonlocal demands of Western Europe. Natural resource utilisation by the East India Company, and later by the colonial rulers, replaced the indigenous organizations for the utilisation of natural resources, like water, forest and minerals, that were mainly managed as commons.

With the establishment of British colonial rule in India, the ever increasing resource demands of-the industrial revolution in England were largely met from colonies like India. Forced cultivation of indigo in Bengal and Bihar, cultivation of cotton in Gujarat and the Deccan led to large-scale commitment of land for the supply of raw materials for the British textile industry, the flagbearer of the industrial revolution. Forests in the sensitive mountain ecosystems like the Western Ghats or the Himalayas were felled to build battleships, or to meet the requirements of the expanding railway network.

Forests of the Bengal-Bihar-Orissa region were used for running wood fuel locomotives in the early stages of railway expansion. The latter stages of colonial resource utilisation and control included the monopolization. of water rights as in the Sambhar Lake of Rajasthan or the Damoda' Canal in Bengal. Colonial intervention in natural resource management in India led to conflicts over vital renewable natural resources like water or forests and induced new forms of poverty and deprivation. Changes in resource endowments and entitlements introduced by the British came into conflict with the local people's age old rights and practices related to natural resource utilisation As a result local responses were generated through which people tried to regain and retain control over local natural resources. The indigo Movement in Eastern India, the Deccan Movement for land rights or the forest movement in all forest areas of the country, the Western

Ghats, the Central Indian Hills or the Himalayas, were obvious expressions of protest generated by these newly created conflict's. Conflicts generated by the colonial modes of natural resource exploitation could not, however, grow with a local identity. With the progress of the anti-colonial people's movement at the national level, these local protests merged with the national struggle for independence.

With the collapse of colonial rule internationally, and the emergence of sovereign independent countries in the Third World like India, resolution of these conflicts at the local level became a possibility. While political independence vested the control over natural resources with the Indian state, the colonial institutional framework for natural resource management did not change in essence. Where colonialism collapsed, the slogan of economic development stepped in.

There was unfortunately no alternative institutional mechanism other than that of the classical model of development left by the British, with which the newly formed Indian state could respond to the accentuated aspirations of the Indian people for a better life. The same institutions and concepts, nurtured and developed by the colonial rulers were applied to objectives which were exactly opposite to those of the colonial period. Concepts and categories relating to economic development and natural resource utilisation that had emerged in the specific context of capitalist growth and industrialization in the centres of colonial power were raised to the level of universal assumptions and applicability.

The processes which led to deprivation were now entrusted with the responsibility of basic needs satisfaction. No serious thought was given to the fact that the historical specificity of early industrial development in Western Europe necessitated the permanent occupation of the colonies and the undermining of the local 'natural economy'.'

This inexorable logic of resource exploitation, exhaustion and alienation integral to the classical model of economic development based on resource intensive technologies led Gandhi to seek an alternate path of development for India when he wrote:

God forbid that India should ever take to industrialism after the manner of the West. The economic imperialism of a single tiny island kingdom (England) is today keeping the world in chains. If an entire nation of 300 million took to similar economic exploitation, it would strip the world bare like locusts.

While Gandhi's critique was a forewarning against the problems likely to arise by following the classical path of resource-intensive development, at the time of India's independence, there was no clear and comprehensive work plan to realise the Gandhian dream of alternate development that would be resource prudent and would satisfy basic needs.

The issues of resource constraints of economic development were, therefore, not highlighted at the theoretical level, partly due to the tremendous pressure of the enhanced developmental aspirations of a newly independent nation, and partly due to the lack of internalization of natural resource parameters within the framework of economics. As the scale of economic development activities escalated from one Five Year Plan to another, the disruption of ecological processes that maintain the productivity of the natural resource base started becoming increasingly apparent.

The classical model of economic development in the case of the newly independent nations resulted in the growth of urban-industrial enclaves where commodity production was concentrated, as well as rapid exhaustion of the internal colonies whose resources supported the enhanced demands of these enclaves.

In the absence of ecologically enlightened resource management methods, the pressure of poverty enhanced the pace of economic development activities in the hope of a quick improvement in the standard of living for all, as in the case of Western Europe. For example, commercial forestry earned more revenue by making increasing amount of timber and pulpwood available in the market but in the process reduced the multipurpose biomass productivity or damaged the hydrology of the forests.

People dependent on non-timber biomass outputs of forests like leaves, twigs, fruits, nuts, medicines and oils were unable to sustain themselves, in the face of the commercial exploitation of forests. The changed hydrological character of the forests affected both the micro-climate and the stream flows, disturbing the hydrological stability and affecting agricultural production.

There are similar examples from all parts of the country, related to almost all massive developmental interventions in India's natural resource system. Ecological degradation and economic deprivation generated by the resource insensitivity and intensity of the classical model of development have resulted in environmental conflicts, an understanding of which is imperative for the reorientation of our current development priorities and concepts.

It is becoming in creasingly clear that these classical concepts and priorities are being used as an alibi to direct 'development' at the national level, while the educated minority elite is the main beneficiary of these 'development' processes.

The ecology movements that have emerged as major social movements in many parts of India are making visible many invisible externalities and pressing for their internalisation in the economic evaluation of the elite-oriented development process.

In the context of a limited resource base and unlimited development aspirations, ecology movements have initiated a new political struggle for safeguarding the interests and survival of the poor, the marginalised, including women, tribals and poor peasants.

5

ENVIRONMENT AND SUSTAINABLE DEVELOPMENT

INTRODUCTION

The purpose of this article is to explore how environmental considerations were reframed so as to become compatible with global development. Adopting an international political economy perspective and based on interviews with the main categories of actors involved, it provides evidence that environmental concerns were remodeled by the joint action of technocratic environmentalists, the international UN-related development establishment and business and industry sectors. Analyzing the results of international cooperation and in particular the review of UNCED's implementation five years after the Summit, the article questions the nature of the "sustainable development" consensus. The inability of the international community to deal with most global environmental issues reveals the limits of international cooperation in the name of the environment economics.

A significant feature of international politics since the end of the 1980s has been the growing concern with environmental protection and the multiplication of the number of international conferences and agreements in this area. Environmental protection is presently recognized as a major political issue, and has acquired a well-defined position on the international political agenda. The United Nations Conference on Environment and Development (UNCED) held in Rio de Janeiro from 3 to 14 June 1992, was a unique moment in diplomatic history. The conference heralded the most elaborate attempt ever to develop institutional solutions to major environmental problems.

Based on the idea that "environment" and "development" had to be linked in a comprehensive framework that would allow for the generalization of economic growth and prosperity while including environmental concerns, UNCED came out with a global solution to the ecological crisis, the concept of "sustainable development." A global bargain was struck, according to which developed nations would provide some financial resources and transfer appropriate and "clean" technology to developing countries to help them protect their environments. An international mechanism - the Global Environment Facility (GEF) - was settled on to undertake the funding of international projects.

The concept of sustainable development is now considered a guiding principle of national and international action. Yet the widespread acceptance of this concept stands in contrast with the inability so far to alter effectively the development model responsible for environmental degradation.

The lack of many positive and concrete results produced by massive efforts in the field of international cooperation for the environment indicate the contradictory character of this new "global" environmentalism. At the same time, global conventions on Climate Change and Biological Diversity were negotiated in an attempt to control the most devastating effects of economic activities, such as CO emissions from industry and consumers, and to protect the earth's living capacity.

A program of action, "Agenda 21," was carefully worked out, covering all areas from health to institutions, from the role of women to the responsibilities of business, all in order to serve as a guide for action to attain sustainability in every country. To facilitate the transition towards "sustainable development," developed countries promised large sums of money in the form of aid, investment and pollution control projects.

The Conference generated a high degree of optimism as to the international community's ability to deal with global environmental problems. Development could continue, now on a truly global base, without the risk of the complete exhaustion of natural resources or of other major environmental catastrophes. The Cold War was over, and rational planning, technology and economic instruments would ensure the extension of the capitalist model of accumulation worldwide.

MAINSTREAMING THE ENVIRONMENT

Global Ecology, International Institutions and the Crisis of Environmental Governance Five years later, at the June 1997 Special Session of the United Nations General Assembly dedicated to the review of UNCED's implementation, the climate was rather different. Optimism had given way to disappointment and, in some cases, there was real concern about the viability of the "sustainable development" model, which relies on a framework of action that does not fully address the causes of environmental destruction. Developed countries have been unable or unwilling to stick to their promise of increasing the aid to development to 0.7% of GDP, as agreed in Rio. Countries like the United States, the largest contributor to global warming, have not shown the will to take effective action that would show a real commitment to reduce their industrial emissions.

On the other hand, developing countries refused to take any further steps without the guarantee that substantive financial resources would back them or that at least the commitments taken in Rio would be respected. The New York 1997 Declaration even recognized that the situation of the environment had deteriorated over the intervening five years, hoping modestly that more progress would be achieved by the next summit in 2002.

The meager positive results produced by the massive efforts in the field of international cooperation for the environment seem to indicate the contradictory character of this new, global "environmentalism."

The purpose of this article is to demonstrate that, while originally being the potential source of a radical and transformative project, environmental concerns were ultimately reframed by the joint action of technocratic environmentalists, the international UN-related establishment and business and industry sectors to become compatible with global development. Adopting an international political economy perspective, the article explores the interaction between state and markets in the construction of global environmental politics. It provides evidence that although there is a new consensus on the diagnosis of the problem - worldwide environmental degradation - very few commitments have been taken to alter the accumulation model and the patterns of production and consumption that contribute to this situation.

It suggests that the failure of the international system in ensuring a move towards sustainability, exemplified in New York, is linked to the very nature of the global bargain struck in Rio. By aiming to make "development" - in its more recent global phase, with its focus on globalized and ever expanding production, trade and consumption - become "sustainable," the concept of sustainability has been stripped of most of its meaning. The inability of the international community to deal with most global environmental issues reveals the contradictory nature of the "sustainable development" consensus and demonstrates the limits of international cooperation in the name of the environment.

Origins and Dimensions of the Ecological Project

In order to understand the meaning of the transformation of environmental concerns into a widely accepted concept, it is useful to recall the original purpose of the ecological project. The ecological movement finds its origins in a protest aimed at defending the right of individuals to regain influence over their ways of living, of producing, and of consuming. As stressed by Gorz (1992), it started as a radical cultural movement, as an attempt by individuals to control and understand the consequences of their actions. With the ecological critique, activists hoped to refocus attention on local knowledge and practices and to bridge the separation of humans from nature, a division that had been at the heart of the Enlightenment project.

In the 1970s, the ecological movement became a political movement, and there was an awareness that the demands of ecology were not only sectorial and local aspirations but rather represented a value shared across national divides (Smith 1996; Gorz 1992).

The publication of the report "Limits to Growth" by the Club of Rome in 1972 gave a scientific backing to these cultural demands and showed the risks posed by the model of industrial growth on the future of life on earth. The report provided a holistic view of the interrelationship between population growth, food production and consumption, the industrialization process, depletion of nonrenewable resources and waste and pollution at the global level, recognizing that waste and pollution are not only a problem for the living conditions and consumption patterns of the population, but affect the very basis of the productive sphere's reproduction.

For the first time, environmental degradation provoked by economic growth was considered from a global perspective, going beyond the occasional questioning of pollution problems during the 1950s and 1960s. In addition, the report launched a real debate on the morality of growth and of the differences in consumption and living standards between developed and developing countries. The 1970s also represented an inflection in the history of social mobilization and collective action with the emergence of the "new social movements," which identify themselves as value movements carrying universal interests going beyond class, nation, sex and race borders.

The new social movements such as the environmental movement appear as "modern" in the sense that they are based upon the belief that history's course can be changed by social actors and are not determined by what Touraine calls a "metasocial principle" (Offe 1988, 219). Environmentalists believe that, although representing a real challenge to our present lifestyles and habits, it is possible to move towards a sustainable society that respects nature and privileges well-being over accumulation. Speaking about the existence of a unique and unified "green movement" is clearly incorrect. Environmental concerns mean different things to different people, take many forms and are expressed through different channels.

In addition, environmentalism takes very different forms in developed or in developing countries. It can mean fighting for an even better quality of life in advanced countries, and fighting for subsistence or even survival in poor countries. Despite this diversity, for the purpose of academic inquiry, three main components of the "green movement," albeit sometimes overlapping, can be distinguished. These three categories should be viewed as "ideal-typical" and not necessarily mutually exclusive.

The first tendency of the ecological movement, deep ecology, is typically a postmodern movement.

In philosophical terms, deep ecology challenges the separation between humans and nature that was at the heart of modern humanism. Deep ecology is not "anthropocentric," it is "ecocentric." As observed by Merchant (1992), it seeks a total transformation in science and in worldviews that will lead to the replacement of the mechanistic paradigm (which has dominated the past three hundred years) by an ecological framework of interconnectedness and reciprocity. The ideas of deep ecology have influenced (among others) Greenpeace, the largest green NGO, which claims that humanist value systems must be replaced by supra-humanist values that place any vegetal or animal life in the sphere of legal and moral consideration. Greenpeace is therefore an example of an environmental organization which, based on scientific reports and examinations, acts to change worldviews and consciousness in order to promote a shift to "ecocentrism" rather than trying to act to transform the production systems which lie at the root of environmental problems.

Yet, while having influenced the most well-known environmental NGO, deep ecology remains a fairly marginal wing of the green movement. Deep ecologists have been criticized for their lack of a political critique, failing to recognize that the idea itself of "ecocentrism" is "anthropocentric." As stressed by Merchant, deep ecologists take the character of capitalist democracy for granted rather than submitting it to a critique. Their tendency to refuse to consider economic policy and to assume a purely conservationist standpoint relegates them to a secondary position.

The second component of the "green movement" is what can be called the "social ecology" movement, which is to a large extent composed of people from the "New Left," dissatisfied with Marxism. Contrary to the deep ecologists, social ecologists maintain an anthropocentric perspective: the concern for nature is understood as a concern for the environment of human beings. Social ecologists seek transformations in production and reproduction systems, that is, a transformation of political economy, as the way to achieve sustainability, social equity and well being. Social ecologists see a contradiction between the logic of capitalism and the logic of environmental protection. For them, environmental protection cannot be made dependent upon economic development, because development, in its liberal sense, has meant the subordination of every aspect of social life to the market economy, and can therefore no longer be considered as a desirable goal. The hegemonic view on "sustainable development," which rehabilitates development as the global goal of humans, is thus unsatisfactory. Social ecologists call for a rethinking of the theoretical basis of development that should include not only economic but also political and epistemological dimensions, such as the questions of participation, of empowerment and local knowledge systems.

For them, what makes development "unsustainable" at the global level is the pattern of consumption in rich countries. Thinking about sustainability thus implies considering the contradictions imposed by the structural inequalities of the global system. Finally, social ecologists vary to a certain extent in the North and in the South: generally speaking, organizations in the North sometimes carry their rejection of development as far as to strike postmodern stances, while organizations in the South focus more on equity and on the need to redistribute the benefits of development.

Finally, there is a more technocratic tendency to the green movement, a tendency that tries to make economic growth and environmental protection appear as compatible goals, which need not require a profound change in values, motivations and economic interests of social actors, nor new models of economic accumulation. For them, it is because capitalist production methods and life standards are not developed enough that environmental problems emerge.

The evidence is that environmental standards are higher in richer countries. Technocratic environmentalists seek to preserve the environment through the establishment of international institutions, the use of economic and market instruments and the development of clean and "green" technology. The result is a rather apolitical approach and activists who, though still interested in environmental protection, are not primarily committed to ideas of equity and social justice, or at least not as committed as social ecologists (Gudynas 1993). The technocratic tendency is thus essentially a rich country tendency, although it is also present in some elite circles in the South.

These environmentalists tend to focus on issues of population for example, arguing that the biggest threat to the environment comes from high population growth in the Third World and the pressure it will bring to bear on the stock of natural resources. Technocratic environmentalists usually tend to belong to organizations which have little or no membership, and rely on their technical and legal expertise and on their research and publishing programs to influence decision-making. Through their close relationship with government and other influential actors and their easy access to international organizations, these organizations tend to have a greater impact than activist membership organizations.

Today, it can be said that this technocratic approach appears to be prevailing over both the biocentric (deep ecology) and the social ecology perspectives and has become what is today mainstream environmentalism, which finds its major expression in the concept of "sustainable development." Despite the challenging and radical nature of ecological concerns, the fact that they might present a potential for change in the present economic model, they were ultimately reframed so as to constitute what appears as an apolitical, techno-managerial approach.

The Formation of a Consensus on "Sustainable Development" It is interesting to examine how the apparent consensus around the concept of "sustainable development" was built and how the project of global environmental "management" became hegemonic. Two main actors have contributed to the hegemony of the liberal environmental management project.

One is the scientific and policy-making environmental community, or, in the words of Peter Haas, the environmental "epistemic community" (Haas 1990); the other actor is business and industry.

The Brundtland Report, the United Nations Conference and the Global North-South "Bargain" International environmental politics did not emerge in the 1990s. As early as 1972, a United Nations Conference on the Human Environment took place in Stockholm, launching the era of international environmental negotiations.

Stockholm did produce some significant outcomes, leading to the creation of the United Nations Environment Program (UNEP), based in Nairobi, which coordinates environmental action within the United Nations. The context of the Stockholm Conference was not very favorable to the adoption of strong environmental commitments. Developing countries were unsatisfied with the UN system and preparing the movement for a New International Economic Order.

They were not willing to yield part of their sovereignty over natural resources in the name of environmental protection, and denounced the emergence of "eco-imperialism." The oil crisis of the 1970s relegated environmental protection to a marginal position in international relations. In the 1980s, the international climate started to change as the debt crisis was seriously affecting developing countries and their role and participation in international fora. In this context, "international commissions" were established to try to elaborate global proposals to promote peace and development, such as the Brandt Commission. Efforts were also undertaken to replace environmental protection on the international political agenda. The World Commission on Environment and Development was established in 1983 under the presidency of Gro Harlem Brundtland, and asked to produce a comprehensive report on the situation of the environment at the global level.

The work of the Commission represented a landmark in international initiatives to promote environmental protection as it produced the concept of sustainable development, a concept that would become the basis of environmental politics worldwide. Sustainable development is defined by the Brundtland Report as a development that is "consistent with future as well as present needs" (World Commission on Environment and Development 1987). The concept of sustainable development was built as a political expression of the recognition of the "finiteness" of natural resources and of its potential impact on economic activities. Indeed, the report argues that, while we have in the past been concerned about the impacts of economic growth upon the environment, we are now forced to concern ourselves with the impacts of ecological stress - degradation of soils, water regimes, atmosphere and forests- upon our economic prospects. The report offered a holistic, global vision of today's situation by arguing that the environmental crisis, the developmental crisis and the energetic crisis are all part of the same, global crisis. It offers solutions to this global crisis, which are mainly of two kinds. On the one hand there are solutions based on international cooperation, with the aim of achieving an international economic system committed to growth and the elimination of poverty in the world, able to manage common goods and to provide peace, security, development and environmental protection. On the other hand, come recommendations aiming at institutional and legal change, including measures not only at the domestic level but also at the level of international institutions. The report emphasizes the expansion and improvement of the growth-oriented industrial model of development as the way to solve the global crisis.

The Brundtland Report also promoted the view that global environmental degradation can be seen as a source of economic disruption and political tension, therefore entering the sphere of strategic considerations. For the Brundtland Commission, the traditional forms of national sovereignty are increasingly challenged by the realities of ecological and economic interdependence, especially in the case of shared ecosystems and of "global commons," those parts of the planet that fall outside national jurisdictions. Here, sustainable development can be secured only through international cooperation and agreed regimes for surveillance, development, and management on the common interest.

For example, the consequences of climate change such as rising sea levels and the effects of temperature variations on agricultural production would require deep changes in the economy and impose high costs on all countries, thus leading to very unstable situations. The issue of forest preservation can also fit into this context, since forests contribute to the stability of climate by acting as carbon sinks, and assure the regeneration of ecosystems by providing reservoirs of biological diversity. Preserving forests then becomes more than an ecological concern: it is also a security imperative. So the "environmental security" discourse was also a cause for the need to find a "consensual solution" to issues of environmental protection. The United Nations Conference on Environment and Development (UNCED), held in Rio de Janeiro in June 1992, marked the official institutionalization of environmental issues in the international political agenda.

Twenty years after the 1972 Stockholm Conference, which was on the "Human Environment," Rio meant a real shift in the vision that had dominated environmental politics so far. After Rio, environmental considerations became incorporated into development, and a "global bargain" was struck between North and South on the basis of the acceptance from both sides of the desirability of achieving a truly global economy which would guarantee growth and better environmental records to all. UNCED recognized the "global finiteness" of the world, i.e., the scarcity of natural resources available for development, but adopted the view that, if the planet is to be saved, it will be through more and better development, through environmental management and "eco-efficiency." The UNCED process involved over a hundred and fifty hours of official negotiations spread over two and a half years, including two planning meetings, four Preparatory Committees (Prepcoms), and the final negotiation session at the Rio Summit in June 1992.

The major result of UNCED is called "Agenda 21," a 700-page global plan of action which should guide countries towards sustainability through the 21st century, encompassing virtually every sector affecting environment and development. Besides Agenda 21, UNCED produced two non-binding documents, the "Rio Declaration" and the Forest Principles. In addition, the climate change and the biodiversity conventions, which were negotiated independently of the UNCED process in different fora, were opened for signature during the Rio Summit and are considered as UNCED-related agreements. The "Rio Declaration," which was the subject of much dispute between the Group of 77 (the coalition of developing countries) and industrialized countries, mainly the United States, illustrates well the kind of bargain reached in Rio.

It recognizes the "right of all nations to development" and their sovereignty over their national resources, identifies "common but differentiated responsibility" for the global environment, and emphasizes the need to eradicate poverty, all demands put forward by the Group of 77. In return, the suggestions by the G77 to include consumption patterns in developed countries as the "main cause" of environmental degradation and the call for "new and additional resources and technology transfer on preferential and concessional terms" were rejected by OECD countries.

In the end, on the issue of finance, an institution called the "Global Environment Facility" (GEF) was set, under the joint administration of the World Bank, the United Nations Development Program (UNDP) and the United Nations Environment Program (UNEP), as the only funding mechanism on global environmental issues, and OECD countries committed themselves to achieving a target of 0.7 percent of GNP going to ODA (Overseas Development Assistance) by the year 2000, to help developing countries implement UNCED's decisions.

Despite the failure of the G77 to win significant concessions on financial resources, if one considers the differences in priorities between developed and developing countries and the conflictual character of the negotiation process, UNCED's outcomes were still seen by the international establishment as quite impressive, marking "an important new stage in the longer-term development of national and international norms and institutions needed to meet the challenge of environmentally sustainable development."

A Commission on Sustainable Development (CSD) was established to monitor and report on progress towards implementing UNCED's decisions. In particular, the CSD's stated aims are to enhance international cooperation by rationalizing the intergovernmental decision-making capacity, and to examine progress in the implementation of Agenda 21 at the national, regional and international levels.

After UNCED, environmental considerations were "integrated" at all levels of action. The "sustainable development paradigm," as some authors recognize, is already replacing the "exclusionist paradigm" (i.e., the idea of an infinite supply of natural resources) in some multilateral financial institutions, as well as in some state bureaucracies and in some parliamentary committees. Most economists now acknowledge that natural resources are scarce and have a value that should be internalized in costs and prices.

Organizations such as the European Union made the "integration" of environmental concerns one of their leading policy principles.

Many countries carried out environmental policy reform to implement UNCED's decisions and the Agenda 21. The boundaries of environmental politics were broadened and its links with all other major issues on the international arena, such as trade, investments, debt, transports, for example, were examined. Efforts were also undertaken to improve environmental records of multilateral finance and development institutions. The World Bank, which has a long history of contributing to environmental degradation by financing destructive projects, went through a "greening" process, and now has a "Department of the Environment" which conducts "environmental impact assessments" and imposes "environmental conditionalities" before granting loans. The World Trade Organization has a "Committee on Trade and Environment" (CTE) which is in charge of ensuring that open trade and environmental protection are mutually supportive. All these efforts can be seen, according to Porter and Brown (1996), as part of a longer-term process of evolution toward environmentally sound norms governing trade, finance, management of global commons, and even domestic development patterns. Environmental considerations were then to be introduced in all major international bureaucracies as a dimension to take into consideration in decision-making processes, and as a challenge for global management. To a certain extent, the "technocratic" approach became hegemonic because it best suited the interests of the international development elite as it magnified its managerial responsibilities. In a time when the legitimacy and utility of the United Nations system was being seriously questioned by its idealizer and major financial supporter - the United States - the goal of making environment and development compatible was seized by some UN agencies as an unexpected opportunity to regain credibility, as well as to be granted funds and to hire new staff for recently created units on "trade and environment" or "finance and environment." UNCED provided a new legitimacy to international organizations such as the World Bank or the World Trade Organization and to their bureaucracies, which now try to assume a leading role in "managing the earth." With the promotion of economic growth to a planetary imperative and the rehabilitation of technological progress, both development institutions and organizations and states appeared as legitimate agents to solve global environmental problems.

If international organizations have benefited from the global perspective that emerged from Rio, they have also contributed to mold it. There is an active "epistemic community," which includes both the international organization establishment and large environmental NGOs, promoting the "global environmental management" approach.

These groups tend to believe that their moral views are cosmopolitan and universal, and emphasize the existence of an international society of human beings sharing common moral bonds.

In this kind of "same boat" ideology, environmental concerns tend to be presented as moral imperatives, related neither to political nor to economic advantages. It would be a consensual concern, a sort of universal principle accepted over borders and political boundaries. An example of an institution promoting these ideas is given by the Commission on Global Governance. In the words of the Commission, "we believe that a global civic ethic to guide action within the global neighborhood and leadership infused with that ethic are vital to the quality of global governance. We call for a common commitment to core values that all humanity could uphold. We further believe humanity as a whole will be best served by recognition of a set of common rights and responsibilities."

Part of the Green movement came to support this "same boat ideology" and was incorporated into the epistemic community. Actually, mainstream conservationist environmentalists were fully admitted into the global environmental management establishment, conferring legitimacy to the UNCED process.

NGOs contributed to UNCED to a degree unprecedented in the history of UN negotiations. NGOs lobbied at the official process, participated in Prepcoms and were even admitted in some countries' delegations, a novelty which was rendered possible by resolution 44/228 calling for "relevant non-governmental organizations in consultative status with the Economic and Social Council to contribute to the Conference, as appropriate."

In addition, during UNCED, NGOs organized in Rio a meeting which ran parallel to the official governmental conference. The "Global Forum," which gathered about 30,000 people, represented 760 associations, among participants and visitors, in a sort of "NGO city." During one week, the Global Forum became home to environmentalists and social activists, to Indians and ethnic minorities, and to feminists and homosexual groups, all united to "save the earth." NGOs organized many demonstrations protesting against the modest results of the official summit and elaborated their own agenda for improving environmental protection worldwide.

Yet, in the eyes of some observers, NGO efforts tended to become coopted by larger and richer groups from advanced countries, which had more means, not only financially but also in terms of organizational, scientific and research capacity, to promote their own views (Chatterjee and Finger 1994).

In the end, NGOs decided that they would sign, in Rio, NGOs "treaties" on all the issues being discussed at the

UNCED official meeting. The main activity at the Global Forum was then the "treaty negotiation" process, just like at the official forum, a process which proved to be very disappointing, as the same North-South conflicts that were blocking UNCED tended to separate northern and southern NGOs.

Ultimately, the NGO treaty process was little more than a pantomime of real diplomacy, and ultimately, the treaties agreed upon, negotiated among a couple of dozen NGOs, had a very modest impact on the future of NGO activities.

The representation at the Global Forum was also very unequal, illustrating differences in means between northern NGOs, very present, and southern NGOs. Asian, and above all, African NGOs, were severely under-represented. Differences in associative traditions and language barriers also explain the hegemony of Anglo-Saxon organizations at the Global Forum.

In the end, influential NGOs decided to concentrate their efforts on lobbying the official conference. The Earth Summit in 1992 thus represented a real moment of acceleration for NGO activities, as it allowed some of them to have a better idea of what their counterparts were doing in other parts of the world, and was the base for establishing cooperation projects and partnerships among organizations. Yet while NGO efforts illustrated by the Global Forum aimed at uniting NGOs worldwide, the green movement came out of Rio appearing even weaker and more fragmented, with the polarization between "realist," co-operative NGOs on the one side and "radical," transformative NGOs on the other.

Finally, the "sustainable development" approach also suited the interests of some governments in the Third World which are primarily committed to economic development and sought through UNCED to obtain concessions in financial and technological terms in exchange of their support for environmental management. Some Third World countries are still marked by a "developmentalist" ideology in which economic development comes before all else.

In addition, resource rich countries such as Malaysia, Indonesia, or Brazil, have traditionally had a vision of unending and expanding frontiers, in which land and natural resources are unlimited and no constraints are seen to exist on the use of resources. As a result, they were unwilling to accept the elaboration of international regimes aiming at limiting their sovereignty over the exploitation of natural resources.

The issue of sovereignty had long been a major source of tension during international environmental negotiations. As long ago as the Stockholm Conference in 1972 developing countries had pressed for the inclusion of a specific principle on the topic. Principle 21 of the Stockholm Declaration stated that "States have, in accordance with the Charter of the United Nations and the principles of international law, the sovereign right to exploit their own resources pursuant to their own environmental policies, and the responsibility to ensure that activities within their jurisdiction or control do not cause damage to the environment of other States or areas beyond the limits of national jurisdiction."

The same debate arose when UNCED was convened, and in the end the sovereignty principle as in stood in the Stockholm Declaration's Principle 21 was included in the Rio Declaration.

In addition, a guarantee that economic development would continue to be the priority on the international agenda was an essential element for developing countries. The reaffirmation of the right to development, and of the sovereignty principle, ensured in Rio, were then the two elements that made agreement at UNCED possible for the Group of 77. The alliance between environment and development could then become official. As described by the vice-president of the International Institute for Environment and Development (IIED), "it has not been too difficult to push the environment lobby of the North and the development lobby of the South together. And there is now in fact a blurring of the distinction between the two, so they are coming to have a common consensus around the theme of Sustainable Development" (World Commission on Environment and Development 1987, 64). Yet to fully understand the nature of this consensus around sustainable development, one last actor needs to be introduced. The actor whose vision shaped most fundamentally the content of this consensus and the real winner of Rio, the business and industry sector, and in particular transnational corporations.

THE INFLUENCE OF BUSINESS AND INDUSTRY

Throughout this process of consensus formation, business and industry exerted a structuring influence. They succeeded in making their view hegemonic, and ended up being considered post-Rio as a major social actor providing solutions to the global ecological crisis. As influential economic agents, transnational corporations (TNCs) have activities that directly impact on the situation of the environment. TNCs have been a constant target of NGOs, which point out their preponderant role in environmental degradation. Several public campaigns and boycotts have been organized to draw the public's attention on the issue and force TNCs to comply with legislation, adopt higher environmental standards or change production processes.

On the issue of tropical deforestation for example, NGOs have pointed out that corporations such as British Petroleum, Shell or Mitsubishi bear a large responsibility for forest devastation worldwide. Already in 1989, The Sunday Times directly accused British Petroleum and Shell of contributing to the depletion of the Amazonian rainforest in Brazil.

More recently, the Rainforest Action Network (RAN) accused Mitsubishi, together with its subsidiary Meiwa, of being "the greatest corporate threat to the world's tropical, temperate and boreal forests." RAN accuses Mitsubishi of illegal logging, transfer pricing, tax evasion, violations of pollution standards, anti-trust activity, violation of native land claims, and employment of illegal aliens. Yet despite evidence of the role of corporations in environmental degradation, the issue was scarcely discussed and questioned during the UNCED process. There is, it is true, a chapter in Agenda 21 dedicated to the role of business and industry. Yet the document does not in any way blame business for its major contribution to the ecological crisis. Agenda 21 contents itself with providing guidelines to firms in order to help them improve their environmental records.

But this is not to say that business and industry were absent or uninterested in the negotiation. On the contrary, large corporations were very active in the UNCED process, and even before it. As early as 1984 a World Industry Conference on Environmental Management (WICEM I) had been organized in France to recommend actions to include environmental concerns in industry planning. WICEM II, which took place in 1991, adopted sustainable development as its main axiom.

The corporations agreed that there should be convergence, and not conflict, between economic development and environmental protection, and launched the Business Charter for Sustainable Development. In 1990, the Business Council for Sustainable Development (BCSD) was created under the chair of the Swiss industrialist Stephan Schmidheiny, personal friend of Maurice Strong (UNCED's Secretary General) and his special adviser for business and industry during the UNCED process. The BCSD was created as a group of 48 chief executive officers of corporations from all regions of the world, some of them with a rather negative environmental record, including Chevron, Volkswagen, Nissan, Nippon, Mitsubishi, Dow, Shell, CVRD, Aracruz, and Axel Johnson. The BCSD was closely involved in the preparation of the Conference, and, through Strong, had special access to UNCED's Secretariat. As a result, after Rio, corporations became "partners in dialogue," and their vision of sustainability became the dominant vision. According to Chatterjee and Finger (1994), corporations shaped the very way environment and development are being looked at: business and industry's worldview came out of Rio as the solution to the global environmental crisis and no longer as its cause.

In the words of the BCSD, "the cornerstone of sustainable development is a system of open, competitive markets in which prices are made to reflect costs of environmental as well as other resources. When viewed within the context of sustainable development, environmental concerns become not just a cost of doing business, but a potent source of competitive advantage. Enterprises that embrace the concept can effectively realize the advantages in more efficient processes, improvements in productivity, lower compliance costs, and new market opportunities." Thus, by creating competitive advantages, environmental concerns can provide corporations with new market opportunities and be the source of new profit. Finally, business sees the new era of global development as the era of market efficiency.

"It is time for business to take the lead," says Schmidheiny; "change by business is less painful, more efficient, and cheaper for consumers, for governments, and for business themselves. By living up to its responsibilities, business will be able to shape a reasonable and appropriate path toward sustainable development" (Schmidheiny 1992, 28-30; Chatterjee and Finger 1994, 122-8). The ecological crisis perceived in fact by business not as a real crisis but rather as a set of adverse and controllable side-effects of development. Hence it is to be solved via increased efficiency which is to be achieved not through government regulation, but through open markets with a new concern for internalizing externalities.

Today, the BCSD has become the WBCSD (World Business Council for Sustainable Development), under the chair of Börn Stigsen. It now has 125 members representing companies such as British Petroleum, Ciba Geigy, Nestle, Monsanto and the Western Mining Corporation. The WCSD is said to have led industry input into the UN Commission for Sustainable Development and UNCED's 1997 review, revealing the emergence of corporate environmentalism as a driving force of global environmental management.

As stressed by Karliner (1997), after Rio, global corporate environmentalism has helped build a public image of transnational corporations as the world's responsible global citizens, setting the terms of the debate along lines favorable to their interests. In the process, corporate environmentalism has partially neutralized efforts - ranging from popular environmental movements to intergovernmental treaties and conventions - that pose a threat to their activities. While before Rio the environmental movement used the system to advance its goals, now the system has appropriated the environmental discourse and is using the environmental movement.

This new strategy has meant increased efforts by corporations to increase cooperation with other environmental actors, in particular with the environmental movement. As noted by Bryant and Bailey (1997, 120), TNCs have sought to cultivate links with moderate NGOs in order to neutralize the threat posed to business from environmentalists. Actually, some NGOs today depend on TNCs for financial support. Stauber and Rampton (1995) observe that this process of funding NGOs and cooperating with them is part of a larger attempt to divide-and-conquer the NGO sector by winning support among moderate NGOs while attacking radical NGOs which campaign against TNCs' activities. Moderate NGOs and TNCs became partners in the international environmental establishment and now work together in the system of global environmental governance.

From Rio 92 To New York 97: The Rise And Fall Of "Global Environmental Management" UNCED's Review Five Years after Rio Five years after Rio, as foreseen at UNCED, the review of UNCED's implementation culminated with the June 1997 New York Summit, often referred to as "Earth Summit II."

Earth Summit II's official name is UNGASS, United Nations General Assembly Special Session. During UNGASS, five years of work of the Commission on Sustainable Development (CSD) were presented, including a report by the Secretary-General assessing the progress achieved in the implementation of Agenda 21 and recommendations for future action and priorities.

UNGASS was carried out at the highest level of political representation - Heads of State and Governments - and, as UNGASS itself said, aimed to "re-energize our commitment to further action on goals and objectives set out by the Rio Earth Summit." A new energy was indeed necessary: the main outcome of the meeting was the public recognition of the failure of international efforts to promote long-term sustainability. Yet it only adopted a document, the "Program for the Further Implementation of Agenda 21," and did not produce a political statement or binding commitments needed to reverse unsustainable trends.

The text acknowledges that, five years after UNCED, the state of the global environment has continued to deteriorate, and reviews the situation in all areas of action.

It notes progress in institutional development, international consensus-building, public participation and private sector actions, which have allowed some countries to curb pollution and slow the rate of resource degradation. Yet, overall, trends are worsening, polluting emissions have increased, and marginal progress has been made in addressing unsustainable production and consumption patterns. Inadequate and unsafe water supplies are still aggravating health problems, the situation of fragile ecosystems is still deteriorating, and non-renewable resources are used at an unsustainable rate. Despite progress in material and energy efficiency, the report concludes that overall trends remain unsustainable. The document then reviews progress in all sectors and issues, inter alia, fresh water, oceans and seas, forests, energy, transport and atmosphere. Finally, it recommends means of implementation and adopts a program of work of the CSD for the next five years, with a commitment to ensure that the next comprehensive review of Agenda 21 in 2002 demonstrates greater measurable progress in achieving sustainable development.

Interestingly enough, all these trends are examined within the framework of economic globalization. The very assessment of progress made since UNCED starts by highlighting that the five years elapsed since then have been characterized by the accelerated globalization of interactions among countries in the areas of world trade, foreign direct investment and capital markets. The document recognizes the unevenness of the globalization process, stressing that marginalization and income inequality is increasing in some countries as well as within countries and that unemployment has worsened in many countries.

Yet it is believed that globalization presents new opportunities and challenges. The report notes that a limited number of developing countries have been able to take advantage of those trends, attracting large inflows of external private capital and experiencing significant export-led growth and acceleration of growth in per capita gross domestic product.

The view is thus that all countries could take advantage of the globalization trend. It is not perceived that only a few countries, due to specific conjunctural conditions, including interest rates and the monetary situation for example, can attract the volume of FDI necessary to feed the high growth rates praised in the document. The conceptual link with economic globalization appears as somehow flawed. It is not mentioned that significant export-led growth and the acceleration of growth in per capita GDP, if not controlled by an effective system of environmental protection, might be responsible for the worsening of overall trends for sustainable development.

In addition, though the text perceives unsustainable patterns of production and consumption as the major cause of continued deterioration of the global environment and observes that unsustainable patterns in the industrialized countries continue to aggravate the threats to the environment, only very vague actions and guidelines are adopted to change them, such as recommending the internalization of environmental costs, developing indicators, promoting efficiency, information, technology, and the role of business in shaping more sustainable patterns of consumption.

No binding commitment to deal effectively with consumption patterns or to establish sustainable production and consumption strategies has been adopted, and the role of actors who tend to promote unsustainable production and consumption patterns, such as business, is actually strengthened.

As well as consumption and production patterns, another distorted linkage to structural economic conditions is made with the recognition that as a result of globalization, external factors have become critical in determining the success or failure of developing countries in their national efforts. It is rightly observed that environmental protection can only be promoted through a shift in the international economy and the establishment of a genuine partnership in order to achieve a more equitable global economy. Yet the idea is that the way to make all countries, in particular developing countries, benefit from globalization is through a combination of trade liberalization, economic development and environmental protection. It is believed that the international trading system should have the capacity to further integrate environmental considerations and enhance its contribution to sustainable development, without undermining its open, equitable and non-discriminatory character.

The text limits itself to recommendations to implement the Uruguay Round and promote trade liberalization. The reality of the present international trading system, a system which promotes discrimination against developing countries, consolidates global disparities and supports unsustainable practices not only in terms of consumption and production but also encouraging transport and pollution and shift from traditional cultures, is not seen as contradictory with the goal of long-term sustainability. With respect to transport, the text notes that the transport sector and mobility in general have an essential and positive role to play in economic and social development, and transportation needs will undoubtedly increase.

It also observes that, in the future, transportation is expected to be the major driving force behind a growing world demand for energy. The document accepts that present trends are unsustainable, and adopted recommendations to make transport become more sustainable and mitigate its negative impacts. Yet the document fails to recognize the major cause of transport's expansion, namely, trade liberalization, which encourages production to relocate on the base of a traditional government subsidy to transports or allows for products originating at the other end of the world to be cheaper than products produced a few miles from the consumer. The fact that the whole globalization project is based on the continuity of cheap transport is not discussed.

Generally speaking, UNCED's review was critically received at all levels, being criticized both by diplomats,

NGOs and by the press. Ambassador Razali Ismail of Malaysia noted that the compact achieved at Rio had eroded along with much of the high-profile attention to sustainable development generated by UNCED. And the Earth Negotiations Bulletin, a publication of the International Institute for Sustainable Development, noted that "in 1992 one could scarcely escape the news of UNCED and/or the environment in the media. This is not the case today... In international relations, perceptions are everything, and if UNGASS is ultimately billed as a non-event it will not bode well for the future of sustainable development or the UN in general during this critical time of its reform."

Most of the world's press was unanimous in condemning the failure of the New York Summit. The French newspaper Libération, for example, noted in its article "The Earth Summit goes round in circles" that the New York summit closed on an acknowledgment of impotence.

Not only did the conference show the little progress accomplished in five years, it also failed to commit governments to significant concrete action and to provide means for implementing Agenda 21. No commitment was taken to achieve the goal of 0.7 % of GDP going to ODA, considered necessary to move towards sustainability. Development assistance today does not exceed 0.3% of GDP, on average, and, in the case of the United States, it was only 0.1 % in 1995.

The US was also the target of much criticism for failing to commit to effectively fighting global warming and to accept concrete reductions in levels of greenhouse gas emissions. At the end of the climate negotiations, no legally binding commitments to target and timetables emerged, and the conference only produced a watery compromise to seek satisfactory results at the then forthcoming Kyoto Conference on Climate Change, which took place in December 1997. In short, on most major issues at stake, New York 1997 represented a backwards step in relation to UNCED's outcomes. NGOs speak of a scandalous betrayal of the Rio promises and of an utterly shameful outcome from Earth Summit II.

The reality is that the world has changed since Rio, and this change has a name: globalization. The Rio 1992 bargain was based on the commitment by developed countries to provide increased financial resources through ODA and technology transfer to help developing countries move towards sustainability. The implementation of UNCED's agreement was in a sense made dependent upon this aid. However, since Rio, ODA levels have been declining and the private sector has become the major agent of change. Government spending is being cut and state reforms are being carried out worldwide, often reducing not only ODA but also domestic environmental budgets. At UNGASS 1997, developing countries through the G77 tried to obtain a recommitment from the North to UNCED's bargain, including an increase in financial flows, technology transfer and an international economic system more favorable to developing countries. Yet today, as foreign investment replaces overseas development assistance in amount and frequency, UNCED's bargain seems politically outdated, and, as a result, its implementation appears highly jeopardized.

Finally, at the level of NGOs, the fracture among environmentalists is today stronger than five years ago. True, NGOs did lobby the CSD and try to influence the official negotiation process. Indeed, NGOs achieved unprecedented access to the intergovernmental process, with Greenpeace and the Third World Network being allowed to make speeches before the General Assembly. However, most of them had given up the idea of having a unified position on all environmental matters, and no "Global Forum II" was organized in New York, only an inappropriately named "Global Gathering" took place.

THE LIMITS OF GLOBAL ENVIRONMENTAL MANAGEMENT

Although steps have been taken between Rio 1992 and New York 1997 in the direction of the globalization of environmental protection, the world seems to be further away from sustainability today than it was then. With environmental globalization and the consensual concept of sustainable development, the perception of an ecological crisis, or at least of ecological limits to development, appears to have vanished: acting for environmental protection increasingly tends to be seen as a technical problem, the task of increasing efficiency and of better using resources such as science, technology, information, capital, and institutions. The cause of environmental problems is no longer perceived as linked to industrial development and ever-increasing material accumulation, but it has become the very existence of human beings. Environmental problems are understood as unavoidable, as side effects of human activities, and efforts are then directed at solving these problems.

Indeed, global environmental management and sustainable development can be seen as "problem solving" concepts, in Robert Cox's terminology, as they only represent a strategy to allow the pursuit of present lifestyles and standards.

Following Strange's call for a critical International Political Economy and the need to address the question of "who gets what, how and why," the analysis of the evolution of the system of global environmental management has revealed that it tends to strengthen the mechanisms of exclusion and inequality (Strange 1988). Global environmental management and sustainable development tend to be uneven concepts, as they do not aim at promoting the correction of global disparities. They attempt to offer a universal framework in which the global society is the unit of analysis and a large share of the blame for environmental degradation rests on the Third World. Instead of stressing affluence, over-production and over-consumption in advanced countries as the main causes of environmental degradation, it tends to suggest that problems arise from poverty. Environmental degradation is transformed from a problem of affluence into a problem of poverty. The responsibility is shifted from major polluters and industrialized countries' abuses to all inhabitants of the planet. For Chatterjee and Finger (1994), the only different element in this approach is that development is now looked at from a global perspective, making the development discourse universal.

And the New York Summit represented a step further in that direction, asserting the desirability of the globalization process and underlining its beneficial aspects. It also consolidated the role of business and industry as privileged partners of the United Nations, establishing permanent contact and consultation on environmental issues. The regulatory situation relating to TNCs and business in general has worsened greatly in the past five years (Khor 1997).

Already in 1992, the US government successfully pressured for downsizing the UN Center on Transnational Corporations (UNCTC), which had been set up to monitor the social, economic and environmental impacts of corporate investment in developing countries. Today, the UN is considering cutting a sub-group of the UN Human Rights Commission which addresses the impacts of corporations on a broad spectrum on rights issues. The main international initiatives and institution for establishing guidelines for the behavior of corporations, which could together lay down a code of obligations and rights of TNCs and states, have disappeared.

In their place has come a strong and growing opposite trend to reduce and remove regulations that governments have over corporations, to grant them increased rights and powers, and to reduce the authority of states to control their behavior and operations.

The Uruguay Round for example has already granted far higher standards of intellectual property rights protection to corporations, thus facilitating further their global monopolization of technology and ability to make profit through higher prices. There are also strong pressures from Northern governments at the World Trade Organization to grant foreign companies the right of entry, establishment and national treatment in all WTO member states (Khor 1997).

In addition, the partnership between the UN and global corporations seems to have been further strengthened in the past two years. In part due to the difficult financial and political situation in which the UN finds itself as a result of the US government's refusal to pay the US $1.6 billion it owes, the UN is now openly seeking political and economic support from corporations. At the last Davos Economic Forum in Switzerland, UN Secretary General Kofi Annan called for a human face to the global market and challenged business leaders to adhere to universal values defined by the UN and contribute to global environmental protection, indicating a broader trend of growing UN collaboration with transnational corporations.

Recently, the United Nations Development Program (UNDP) has solicited funds from global corporations with poor records on human rights, labor and the environment, such as Dow Chemical, Citibank and Rio Tinto Plc, in exchange of special UNDP sanctioned logos for use by corporate sponsors. Called the "Global Sustainable Development Facility (GSDF)," the plan calls for corporate sponsors to funnel donations to a separate entity which they will manage. In the words of the UNDP, the GSDF "brings together leading global corporations and the UNDP, to jointly define and implement a new facility to eradicate poverty, create sustainable economic growth and allow the private sector to prosper through the inclusion of two billion new people in the global market economy."

According to the internal memo, sponsors will benefit from the advice and support of UNDP through a special relationship, allowing corporations unprecedented access to UNDP's network of offices, high level governmental contacts and the knock-on effects of its reputation. The plan, revealed through a leaked internal UNDP memo, has been heavily criticized by observers and NGOs, who warn that the interests of global corporations are often at odds with the basic economic and social needs of the world's poor and the values of human rights and environmental protection the UN is meant to protect.

According to Ward Morehouse, President of the US Council on International and Public Affairs, the UN should be monitoring the human rights and environmental impacts of corporations in developing and industrialized nations, not granting special favors... Increasing collaboration will lead to a rcluctance to criticize corporations which are central players in the human rights, environmental and developmental dramas unfolding every day across the globe.

To conclude, the UN and international organizations in general seem to be moving towards the adoption of a market-oriented global model of environmental governance, which sees economic globalization as a positive and integrative process. The key actors of economic globalization, transnational corporations, are taking the leading role and consolidating their influence on the system of international environmental governance. With the adoption of this project of global environmental management, one particular understanding of the world, the one promoted by business and large corporations in Western affluent societies, becomes hegemonic and appears to be universal (Shiva 1993).

Environmental concerns have been incorporated as a mere dimension of the "globalization project," understood in McMichael's definition as "an emerging vision of the world and its resources as a globally organized and managed free trade/free enterprise economy pursued by a largely unaccountable political and economic elite" (McMichael 1996, 300). This project advocates a universalized model of production, of consumption, and thus of dealing with problems of environmental protection resulting from these activities.

By assuming its universality, it tends to marginalize other knowledge and other solutions to problems of environmental protection. Interestingly enough, it was unanimously recognized that the most positive result and follow-up of UNCED was without doubt realized at the micro-level, within the Local Agenda 21 framework.

In an effort to implement Agenda 21 locally, social groups have worked together with local authorities to make sustainable development a reality at the local level, often on a truly participatory basis and reflecting grassroots concern and involvement.

From a critical point of view, global management's failure, exemplified at the New York Summit, was not entirely unexpected. Indeed, the global management approach inspired by business perspectives and propagated by the international development establishment tends to strengthen the globalization process at work today, failing to counter its effects in terms of social exclusion and environmental destruction. It tends to weaken social protection and environmental protection in the name of economic efficiency (McMichael 1996).

It stands at odds with the commitment to social change and to equity that lies at the root of a critical, political economy view of global environmental politics as inserted within the dynamics of economic accumulation and social structures. The question of the ownership of natural resources, for example, is not addressed. However, environmental problems in the South are often linked to problems of resource ownership and equity.

Sustainable Development as defined in Rio and reasserted in New York has been practically translated into technocratic responses to what are in reality political problems.

CONCLUSION

The purpose of this article was to analyze and evaluate the process of "mainstreaming" of environmental concerns. Recalling the radical and transformative origins of the ecological project, it has provided evidence that environmental concerns have been remodeled by the joint action of technocratic environmentalists, the international UN-related development establishment and business and industry sectors.

Examining the results of international cooperation, the article has questioned the nature of the sustainable development consensus, a consensus deeply marked by the growing access and influence of global corporations on UN activities. Today, market-oriented perspectives to environmental problems seem to be prevailing over more transformative views, especially at the international level, within the framework of international organizations and institutional agreements.

The article has suggested that the mainstream approach to sustainable development tends to reduce ecology to a set of managerial practices aiming at resource efficiency and risk management. In doing so, it tends to address a civilizational impasse as a mere technical problem (Sachs 1993). The mainstream approach proposes that environmentalists should operate using the language and the worldview of Western economics in approaching ecological concerns.

Instead of designing cultural and political limits to development, the project of "global environmental management" tends to become part of a technocratic effort to sustain industrial development in the age of economic globalization.

Environmental protection, together with democracy, human rights and free market economics, becomes a universal consensus, a universal consensus which, as Baudrillard remarks, arouses suspicion, since it is about values that have become devalued, values becoming emptied at the very moment of their hegemony (Huysmans 1995). Environmental concerns become just another element in a process leading to global uniformity, a uniformity of cultures, lifestyles, mentalities, but also of relationships with nature (Sachs 1993; Latouche 1996).

This market-oriented agenda may provide a starting point for dealing with global environmental problems.

The documents, which emerged from international environmental negotiations from Rio to New York, replete with inconsistencies, represent a complex mix of disagreements, hopes and compromises. One may concede that conventions and obligations reflect the need for government negotiators to find the minimum agreeable grounds to initiate a large open-ended process on major environmental issues.

In this sense, they only produced a general framework for negotiations, steps on the way to building international regimes. They do not form a series of real commitments representing an effective consensus on how to deal with global environmental issues.

Yet considering the amount of time, energy and resources invested in this process of international environmental regime building, one might have hoped for more concrete, positive results. The failure of the present framework to effectively promote sustainability, which became evident in the 1997 New York summit, is recognized even by one of the major promoters of this path, Maurice Strong, ex-secretary general of UNCED, today President of the Earth Council. For Strong, unfortunately, the economic, social and demographical forces that lead to unsustainable development still prevail. Strong sees the lack of political will from governments as the main cause of this failure.

The present framework appears to contain many contradictions that limit the ability of the international and national communities to solve satisfactorily environmental problems.

Adopting an international political economy perspective, this article has argued that the main problem that international efforts to protect the environment have to address is the issue of the impact of economic globalization.

Economic globalization has in a sense helped to create conditions for the development of policy mechanisms and institutions that will universalize and promote the concept of "sustainable development." Global change is exerting a structuring influence on the redefinition of environmental politics.

However, the kind of sustainable development being promoted seems to represent more the consolidation of a global project of "environmental management" than a real shift away from destructive practices. Globalization is consolidating a market-friendly view of sustainable development, a view that gives priority to the sustainability of "global growth" and to the correction of environmental damage. This tends to be carried out at the expense of the competing alternatives and participative view of sustainable development as stressing not only development but also social equity and decentralized participation.

The "globalization project" has shaped and redefined both the content of environmentalism and environmental policies and structured the international political economy in a way that makes sustainability more difficult to achieve.

6

CONTROVERSIES OF THE ENVIRONMENTAL ECONOMICS

AIR QUALITY STANDARDS REVIEW

Since its inception the EPA has begun to rely less and less on its scientists and more on nonscience personnel. EPA has recently changed their policies regarding limits for ground-level ozone, particulates, sulfur dioxide, nitrogen oxides, carbon monoxide and lead. New policies will minimize scientist interaction with the agency and rely more on policy makers who have minimal scientific knowledge. This new policy has been criticized by Democrats.[9] On March 12, 2008, the Federal government of the United States reported that the air in hundreds of U.S. counties was simply too dirty to breathe, ordering a multibillion-dollar expansion of efforts to clean up smog in cities and towns nationwide.[10][dated info]

FUEL ECONOMY

In July 2005, an EPA report showing that auto companies were using loopholes to produce less fuel-efficient cars was delayed. The report was supposed to be released the day before a controversial energy bill was passed and would have provided backup for those opposed to it, but at the last minute the EPA delayed its release.[11]

The state of California sued the EPA for its refusal to allow California and 16 other states to raise fuel economy standards for new cars.[12] EPA administrator Stephen Johnson has claimed that the EPA is working on its own standards, but this move has been widely considered as an attempt to shield the auto industry from environmental regulation by setting lower standards at the federal level, which would then preempt state laws. California governor Arnold Schwarzenegger, along with governors from 13 other states, stated that the EPA's actions ignored federal law, and that existing California standards (adopted by many states in addition to California) were almost twice as effective as the proposed federal standards.[16]It was reported that Stephen Johnson in making this decision, ignored his own staff.

GLOBAL WARMING

In June 2005, a memo revealed that Philip Cooney, former chief of staff for the White House Council on Environmental Quality, and former lobbyist for the American Petroleum Institute, had personally edited documents, summarizing government research on climate change, before their release. Cooney resigned two days after the memo was published in The New York Times. Cooney said he had been planning to resign for over two years, implying the timing of his resignation was just a coincidence. Specifically, he said he had planned to resign to "spend time with his family."[19] One week after resigning he took a job at Exxon Mobil in their public affairs department.

In December 2007, EPA Administrator Stephen L. Johnson approved a draft of a document that declared that climate change imperiled the public welfare - a decision that would trigger the first national mandatory global-warming regulations. Associate Deputy Administrator Jason Burnett e-mailed the draft to the White House. White House aides - who had long resisted mandatory regulations as a way to address climate change - knew the gist of what Johnson's finding would be, Burnett said. They also knew that once they opened the attachment, it would become a public record, making it controversial and difficult to rescind. So they didn't open it; rather, they called Johnson and asked him to take back the draft. U.S. law clearly stated that the final decision was the EPA administrator's, not President Bush's. Johnson rescinded the draft; in July 2008, he issued a new version which did not state that global warming was danger to public welfare. Burnett resigned in protest.

GREENHOUSE GAS EMISSIONS

The Supreme Court ruled on April 2, 2007 in Massachusetts v. Environmental Protection Agency that the EPA has enough data to make a ruling on whether or not to consider Carbon Dioxide as a green house gas, stating that "greenhouse gases fit well within the Clean Air Act capacious definition of air pollutant."

Jason K. Burnett, former EPA deputy associate administrator, told the United States Congress that an official from Vice President Dick Cheney's office censored congressional testimony by Julie L. Gerberding, director of the Centers for Disease Control and Prevention. Reportedly, the testimony excluded said that "CDC considers climate change a serious public health concern."

LIBRARIES

In 2004, the Agency began a strategic planning exercise to develop plans for a more virtual approach to library services. The effort was curtailed in July 2005 when the Agency proposed a $2.5 million cut in its 2007 budget for libraries. Based on the proposed 2007 budget, the EPA posted a notice to the Federal Register, September 20, 2006 that EPA Headquarters Library would close its doors to walk-in patrons and visitors on October 1, 2006.[24] The EPA also closed some of its regional libraries and reduced hours in others,[using the same FY 2007 proposed budget numbers.

On October 1, 2008, the Agency re-opencd regional libraries in Chicago, Dallas and Kansas City and the library at its Headquarters in Washington, DC.

MERCURY EMISSIONS

In March 2005, nine states, California, New York, New Jersey, New Hampshire, Massachusetts, Maine, Connecticut, New Mexico and Vermont, sued the EPA. The EPA's inspector general had determined that the EPA's regulation of mercury emissions did not follow the Clean Air Act, and that the regulations were influenced by top political appointees.[27][28] The EPA had suppressed a study it commissioned by Harvard University which contradicted its position on mercury controls. The suit alleges that the EPA's rule allowing exemption from "maximum available control technology" was illegal, and additionally charged that the EPA's system of pollution credit trading allows power plants to forego reducing mercury emissions. Several states also began to enact their own mercury emission regulations. Illinois' proposed rule would have reduced mercury emissions from power plants by an average of 90% by 2009.

Air Ratings

An Environmental Protection Agency employee checks one of the many air sampling locations set up around the World Trade Center site.

VERY FINE AIRBORNE PARTICULATES

Tiny particles, under 2.5 micrometres, are attributed to health and mortality concerns so some health advocates want EPA to regulate it. The science may be in its infancy although many conferences have discussed the trails of this airborne matter in the air. Foreign governments like Australia and most EU states have addressed this issue.

The EPA first established standards in 1997, and strengthened them in 2006. As with other standards, regulation and enforcement of the $PM_{2.5}$ standards is the responsibility of the state governments, through State Implementation Plans.

SYSTEMS OF WATER MANAGEMENT

Large-scale water projects which work against nature's water economy and people's sustenance requirements have been designed by destroying water technologies which were ecologically more sustainable and socially more just.

The tank system of South India is among the indigenous alternatives which has survived over centuries. This system consists of a single series of several hundred and in some cases over a thousand reservoirs linked together and forming such continuous chains of works that not a single drop of water falling in the catchment is lost in times of drought, and very little is lost during normal periods. Major Sankey, one of the first engineers of Mysore State, who concentrated on the systematic repairs of tanks, stated that 'to such an extent has the principle of storage been followed that it would require some ingenuity to discover a site within this great area for a new tank'.

These tanks play a central role in irrigation even today. In the Ravalseema region. in the southern part of the Krishna basin, tanks irrigate 620,000 acres while major and minor irrigation projects cover 427,000 acres. In the Anantapur region, river water was diverted with the help of sand dams. These sand dams were constructed in areas like Koppalakonda, Penakacharala, Kalluru, Tarimala and Rachepalli of Anantapur taluk, Panidi, Appeyipeta, Naganapuram and Chitrachedu of Gooty taluk. In places like Ramapuram, Kallavapalle-and Budigamma, the surface flows in the Pennar were diverted through masonry dams called 'Panthams'. In Hospet, Hagari, Rayadurge and Kudligi taluks, channels were constructed to draw water from rivers for irrigation. Similar types of irrigation schemes have also been reported in Dharwar, Bijapur, Sholapur, Satara, Sangli and Ahmednagar districts. Two such major schemes were irrigating nearly 580 acres in Bijapur district. In Madras Presidency, irrigation through small tanks and canals, which the villagers managed themselves, collectively irrigated an area equal to that irrigated by all the larger works which have been constructed by the British government in that Presidency.

An example of indigenous water use was the widespread system of 'Ahars' and 'Pyres' used for irrigation of paddy fields in South Bihar as reported by Sengupta.4 An 'Ahar' is constructed by erecting an embankment 1 or 2 metres in height on the lower ground. From the two extremes of this embankment two other embankments are constructed so as to project towards the higher ground, gradually diminishing in height as the ground level rises and ultimately ending at the ground level. 'Ahars' were built on drainage rivulets to collect water. Constructions with sides more than a kilometre in length and irrigating over a thousand acres of land were frequently found. 'Pyres' on the other hand were systems devised for utilising the water which flows through hilly rivers running from south to north and intersecting the whole country. 'Pyres' were laid off from the rivers to carry water to-agricultural fields. Some of the largest 'Pyres' were 20 to 30 km in length, fielding a number of distributaries and irrigating may be 100 villages.

The system protected agriculture to such an extent that when there was a famine in other parts of the country, Gaya district, where 'Ahars' and 'Pyres' were most evolved, was not affected. However, this immunity was eroded following the deterioration of these irrigation works, which was primarily a consequence of breakdown of economic and social systems which ensured maintenance of the water systems.

The irrigation works in pre-British India were managed by a variety of social organizations within the village community. Usually, the structure of such organisations included a collective of all the beneficiaries of irrigation works, and was headed by a leader. Based on the leadership, the system was known by different names in various regions.

In Maharashtra districts of the basin, the bandharas were managed by such water committees. The task of these committees was to maintain the earthern dam or diversion from the river and to desilt the canals. Similar work was performed by committees in Bijapur, Dharwar, Raichur and Bellary districts of Karnataka. Similar committees were also found in the districts of Andhra Pradesh where there were known as 'pinnapeddandarule' or 'peddandarule' system. The difference between the two was in terms of the members who constituted them. In the 'pinnapeddandarule' system, youth were preferred as the desilting of canals involved hard physical labour. In Krishna district, less labour intensive work was involved hence the membership rules were flexible. Work involved in desilting or erecting canal banks or rebuilding canals was equally shared by all the beneficiaries. Each member or beneficiary was supposed to do his share of work in proportion to the land held. The committee levied a fine on any one who failed to do his share of work, irrespective of the size of his landholding.

Similarly in South Bihar, both the construction and maintenance of water systems was collectively managed. Each cultivator had to contribute labour to a collective system called 'goam' to repair embankments and desilt channels. In South India this practice was known as 'kudimarammath'.

Allocation of water within the villages was also managed by the cultivators themselves. A system known as 'parabandi' was in operation which regulated the distribution of water among the villages from a common source to ensure fair distribution to all the villages. In case of some large works, the rights of each village were formally recorded. In others, the regulations were largely customary. If conflicts arose, they were resolved according to local rights and regulations.

When the British arrived in India, they had no expertise in water management, since agriculture in Europe has never been dependent on irrigation. Arthus Cotton, the founder of modern irrigation programmes wrote:

There are multitudes of old native works in various parts of India. These are noble works, and show both boldness and engineering layout. They have stood for hundreds of years....

When I first arrived in India, the contempt with which the natives firstly spoke of us on account of this neglect of material improvements was very striking, they used to say we were a kind of civilized savages, wonderfully expert about fighting, but so inferior to their great men that we would not even keep in repair the works they have constructed much less even imitate them in extending the system

Greenway, while writing on farming in India, commented on the lack of advanced engineering knowledge of the British which had led to the catastrophe in Sheffield because of the Bradfield dam. A comparison naturally presents itself between the dam of the Bradfield reservoir, which failed, and the Indian model which has been so long and in so many instances successful, and which if rightly constructed and faithfully attended to, may be regarded as ensuring the maximum of efficiency and safety.

Empirical as the process may appear, practice has made it perfect. Engaged in a continual struggle with the powers of nature, contending with volumes of water far larger, floods far heavier, tempests more violent, than any known in England, the Indian engineers have been forced into devising means, not only to enable a bank to stand a given presence of water, but also to provide resources against contingent risks and accidents, which latter provision, strange to say, appears from the evidence on the inquest of Sheffield not invariably to be considered a part of the engineer's duty.'

The British, however, had to control river water in India keeping in view their economic interest in increasing revenue. In Rajasthan they controlled water to maximise the revenues derived from their monopoly in salt trade, in Bengal to protect their transport network and increase revenues from agriculture. Water politics was thus as transient as the resource itself, to create obey areas and command areas, i.e., areas and people inhabiting them which are callously devastated to cater to the needs of areas and people which are industrialised, urbanised, and politically and economically powerful.

Langdon Winner has suggested that politics is embedded in artifacts. Large water projects definitely have politics built into them. In an attempt to control rivers, they also control the lives of those who depend on the river and its basin for livelihood. Worster has proposed the thesis that making more demands on the earth and devising the means to fulfil them leads to an unequal power division in society. 'Intensification of use eventually must give rise to potent anti-democratic forces whatever their guise may. Karl Wittfogel's idea of a hydraulic society captures the reality of the social organisation created by large dams. Wittfogel's theory is one of power, he holds that control over water leads to control over people.

Wittfogel, however, like Marx before him, perceived the hydraulic society as a reality in Asian civilizations over thousands of years. They assumed that the pervasiveness of and decentralized network irrigation systems was linked to centralized power and that specific individuals conquering river waters turned into a power elite. What Marx and Wittfogel failed to perceive from a distance was that irrigation systems in India were managed by cooperating communities, not dominating bureaucracies, and that decentralized maintenance and use rather than centralised control was the characteristic of India's ancient irrigation systems.

Sengupta has challenged Wittfogel's thesis that domination was characteristic of Asia's water systems. Vast networks of irrigation systems are not necessarily large projects. They can be a closely knit network of micro-projects, each managed locally in terms of construction, maintenance, allocation and conflict resolution among users. Further, stagnation has not been a characteristic of these traditional irrigation systems. Instead, flexibility was often displayed.

The cropping patterns were changed annually according to the availability of irrigation water in a particular year. With water resources under local control, local decisions on land use involved less risks and higher certainty. On the other hand modern canal irrigation from large dams centralises water control and distribution. The time of water supply rarely corresponds to the time of requirement and certainty is far less than it would have been had the cultivators themselves distributed the same amount of water through local resources. Correspondingly, they have much less scope to alter their cropping and irrigation practices to suit the availability of water.

Centralised societal domination is, therefore, linked to centralised water control not to vast networks of decentralised waterworks. The power of the modern state over people as exemplified through large dams is qualitatively different from the social organisation of indigenous irrigation systems. Such political control does not merely violate human rights in the present, it threatens to deny future generations the right to life-support systems. Large water projects interfere in a major way with the natural flow of water and the hydrological cycle. Ecological hazards are intrinsically associated with them, which in turn generate another level of conflicts over natural resources. Social control over water use in indigenous systems had prevented both over-use and abuse of water, and avoided a conflict between the use of water for human consumption and its functions in the maintenance of essential ecological processes, central to which is the water cycle.

COLONIAL CONFLICTS OVER RIVER WATERS

Under colonial influence, water was diverted from its role in the survival economy and nature's economy and was transformed into a source of revenue and taxes, or as an input to commodity production for the generation of profits. The introduction of market forces in the water economy of the country created new conflicts over water resources between the market and survival economies.

The Salt Economy and Water Conflicts

On 5 April 1930 Mahatma Gandhi launched the national noncooperation movement with the campaign for the production and distribution of salt at Dandi beach in Gujarat, violating the Salt Law of the British that had guaranteed their monopoly in the production and distribution of salt. Salt is a vital resource for the survival of both human and animal life specially in tropical countries like India. It remained a common resource till the British monopolised its production and distribution to transform it into a source of revenue.

The growth imperative compelled the expansion of the salt industry, as monopolised by the British, at the cost of diversion of resources from essential economic activities like food production. Further, in order to increase revenue the British government raised the salt tax in 1923 through the Indian Finance Bill of the Viceroy. This triggered off strong protests all over the country since a basic resource like salt was being denied to the people in order to increase revenue. The anguish of Gandhi on this appropriation of a common resource was clear when he said,

They even tax our salt-a necessity of life, only less necessary than air and water. It ought to be free as they are.... Nature bestows it on us and we may not use it. There is salt beside the sea and they forbid us to gather it's

The production of salt in certain pockets like Rajasthan and in the coastal areas as well as its distribution in all parts of the country had been a part of the living history of the Indian civilisation. Salt is one of those commodities which cannot be produced everywhere notwithstanding the self-sufficiency of the village economies. The demand for salt from all parts of the country led to the establishment of salt industries wherever it was possible to extract salt from saline water. All along the coastline except in Bengal, salt was manufactured by digging hollows in the ground and allowing the saline water to evaporate.

The Sambhar Lake in Rajasthan was a major inland source of salt for the whole of Central India. The manufacture and trade of this essential commodity and the possibility of a high margin of profits in case of monopoly in production, led to the East India Company securing the entire lake. In 1870 the British government forced a treaty on the States of Jodhpur and Jaipur through which it acquired the right of not only manufacturing and selling salt but also levying a duty on it. The peculiarity of the Sambhar Lake is that there is no rock salt bed in the geological formation of the area. The salt granules in the great salt marsh of Kutch are carried north-east by the trade winds in the summer and deposited in the catchment of the Sambhar Lake. In the monsoons the salt granules are dissolved and the solution is carried into the lake by a large number of streams. the largest of which is Rupnagar.

Through the quick evaporation of water from the lake. millions of tons of salt get deposited in the Sambhar Lake. According to the British estimates of that period. 46 million people in India depended on supplied from this lake. This made the control of this 90 sq miles lake economically valuable for the growth of the salt industry

Though the British salt industry in India grew by more than 33 per cent in the first twenty years of its existence, it was not content with only monopoly in production and distribution. To enhance the profits and growth of the industry, the British had maximised the inflow of water into the Sambhar Lake by denying the more basic uses of water for irrigation and drinking.

Rupnagar, the largest stream feeding the lake, was the main source of water for the State of Kishengarh whose most important economic activity was agriculture which depended solely on the water of Rupnagar for irrigation. Thus the British attempt to increase the inflow of water into the lake came into direct conflict with the survival needs of the people in Kishengarh.

In 1900, a famine year, Kishengarh wanted to store some water by constructing an embankment tank on Rupnagar. The salt authorities strongly opposed this use of water for drinking and irrigation by the people. The Finance and Commerce Department of the then Government of India promptly accorded higher priority to the interest of the salt industry as opposed to the water needs for the survival of the people of Kishengarh. In a note dated 13 July 1901 the Department issued the following statement:

In view of the way in which the manufacture of salt depends on a sufficient supply of water in the lake and the precariousness of the supply, the Govt. of India consider that it is most inadvisable that anything should be done in the shape of constructing new reservoirs, irrigation works, or of extending any existing works on any of the feeder streams of the lake, either in the British territory or in native states which will be likely to diminish supply....

The Government of India desires that in future the Salt Commissioner may be consulted before any of the existing works' in British territory or the native states are enlarged, strengthened or improved.

In 1906 the villages of Kishengarh faced a severe drought. In sheer desperation to ensure a minimum supply of water for the biological survival of human and animal life, the people built an earthen dam which the British promptly destroyed. In the same year the rulers of Kishengarh pleaded for permission to construct low masonry weirs across Rupnagar and to sink some wells. The Salt Commissioner made sure that these projects could not be undertaken. It is this undisclosed destruction of agricultural ecosystems in the catchment of the Sambhar Lake that made possible and visible the growth of the salt industry there. By 1922 the prosperous and populated Kishengarh was desolate. Large areas of cultivable land became waste and wells could no longer be used. Another desperate attempt by the kingdom of Kishengarh to construct four irrigation tanks was postponed for twenty-four years, thus ensuring the total destruction of the local agricultural economy.

This was the general pattern of destruction of the survival base of the people in all regions of the country where the salt monopoly led to economic growth for British interests. This monopoly forced the people to give up the production of a basic resource like salt, which was being produced by them with simple technologies that required no capital input. The control made people totally dependent on the British supply, to be bought at a price determined by British interests. It was this exploitation by the British through the denial of a basic need as well as the associated destruction of other related resources that led to non-violent violation of the Salt Law as an assertion of the people's right to vital natural resources for survival.

As a mode of protest it spread throughout the country with people breaking the Salt Law in large numbers. Its impact, both immediate and long-term was immense. By 5

March 1931 the British government was compelled to retrace its steps and issued a notification that people could produce salt for their own use.

THE DAMODAR CANAL AND WATER CONFLICTS

Burdwan district of Bengal lies in the Damodar river basin. Irrigation in this district, as in other parts of Bengal, was based on the overflow irrigation system. This ancient system of water use had met the needs of rural Pengal for more than 2,000 years. It provided a single solution to Bengal's inter-locking problems of flood and malaria control, cheap inland transport for the better part of the year, and renewal of soil fertility. This ecological management of common resources led to the establishment of a community organization based on irrigation rights which required every individual to regard his neighbour's interest as his own.

Being a tropical monsoon region, Bengal receives heavy rainfall in the wet season. But due to high percolation rates, the groundwater level declines rapidly after the rains and a serious shortage of water occurs at a time when it is most needed. As a result tanks and wells dry up. The only solution is to maintain saturation of the subsoil by impounding as much rainfall as possible on the surface and keeping it there for as long as possible. Earlier this was done by flooding the tanks and rice fields with the muddy spill of the rivers in spate and by storing rain water in tanks which at one time numbered over 50,000 in Burdwan district alone. Percolation from rice fields and storage tanks maintained the underground reserve of water at a level 10 to 15 feet higher than that at present and prevented the drying up of wet crops in summer and of wells and tanks in the dry season. A system of inundation canals was in operation in Bengal from ancient times. The silt bearing top film of water from flooded rivers was allowed to flow into the rice fields, where the silt was deposited. River silt is very high in manure. It enriched the crops as well as killed noxious weeds in tanks and in the rice fields. Every tank had its distributary to flush it, and every rice field had its distributary to irrigate it. The silt also carried carp eggs. Carp, being larvicidal fish, devour the larvae of the anapholis, thus controlling malaria and at the same time providing the much needed nourishment to the rice-eating peasantry. Removal of silt from the river water prevented the silting up of the rivers at the mouth so that there were no floods.

The political destabilisation of Bengal led to the ecological disruption of this highly efficient system. The water channels were silted up and were declared to be 'dead' or 'blind' by ignorant British engineers. A series of devastating floods followed and by treating the symptoms instead of the cause, the problem was further aggravated. The erection of embankments to protect the Eastern Bengal Railway cut across the natural contours of land, disrupting the natural drainage, causing severe ecological problems and hardships to the once prosperous peasantry.

At the end of the eighteenth century the British took over the responsibilities and liabilities of maintaining the irrigation system with its water resources, spill channels, streams dykes, pools, tanks and embankments. However, the government failed to keep its promise and worked in the opposite direction by making the left embankment watertight. As a result, its innumerable spill channels were closed, and subsequently, the zamindars and tenants made a number of secret breaches through the embankment. In the period between 185-59 the government cut off 20 miles of embankment on the right side of the Damodar with a view to protecting the Railways and the Grand Trunk Road. This led to silting up of the river bed, death of the live channels and stagnation of water due to disruption of natural drainage. This was followed by severe epidemic of malaria which wiped out one-third of the total population in a district in a decade. Areas that were once famous as 'health resorts' became 'decadent areas'.

Sir William Willcocks, one of the greatest irrigation engineers the world has ever known, pleaded in vain for some funds to resuscitate these 'dead rivers' with the cooperation of the peasantry. Once the vicious circle of destruction was broken, the project would run itself and pay its way. But the irrigation and survival needs of the peasantry came into conflict with the interests of railway transport and stream navigation river flotilla companies. The use of water that could have restored to the peasantry'the old prosperous days when irrigation with the muddy water of the Ganges flood was the heritage of all'" was not attempted because spillways and railways were mutually exclusive in the Burdwan tract. Millions were spent on the externalities of the watertight embankments, malaria and flood control projects. For seventy years, embankments were allowed 'to impoverish lands, and impoverish people and affect them with malaria, when a trifling expenditure of the money could bring relief'. In the words of Sir William Willcocks,

The irrigation Department has tried its hand at every kind of irrigation except the ancient irrigation. The resulting poverty of soil, destruction of fish, introduction of malaria and congestion of the rivers have stalled the canals and banks, and the country is strewn today with the wrecks of useless and harmful works.

To compensate the people of Burdwan the government constructed the Eden Canal in 1881 and the Damodar Canal in 1933. The latter was intended to irrigate 20,000 acres of rice lands of 379 villages at the cost of Rs. 1.25 crores. After the inauguration of the canal, the government planned to realise a part of the capital expenditure by imposing a canal tax on the ryots. As the tax was heavy, the peasants refused to execute any lease in order to use the canal .vater. As a result the government introduced a legislation to impose a compulsory levy. The Bengal Development Bill 1935 was introduced by Khwaja Nazamuddin on 18 February 1935 which provided for the improvement of land in Bengal and the imposition of a levy in respect of increased profit resulting from the improvement works constructed by the government. The aim was to induce people to use canal water and to convert this consumption into a source of revenue. The Bill, it was argued, would not only compel the ryots to pay up to half of their increased profits, but would also enable them to make increased profits by taking advantage of the improvements. The Minister stated that peasants did not use canal water in a normal rainfall year but viewed irrigation as an insurance against the failure of monsoon. The Bill was aimed at 'not allowing any man to indulge in the luxury of not consuming canal water during normal rainfall years by making every man pay an improvement levy of Rs. 5.80 per acre per annum irrespective of the benefits derived or likely to be derived from the irrigation facilities of the canal.

The Bengal Development ' Bill was obviously resented by the Burdwan peasantry for whom it became yet another source of colonial exploitation. In the later thirties the popular discontent gathered momentum in the command of the Damodar Canal over the Bengal Development Act and the improvement levy. A movement crystallized as the Damodar Canal Tax movement. An association called the Burdwan District Raiyats Association was formed with D.P. Choudhury and Balai Chand Mukhopadhyay as President and Secretary, respectively to oppose the Act and the canal tax.

On 20 December 1935, peasants from 500 villages of the Damodar Canal area held a mass meeting under the auspicious of the Raiyats Association. The meeting adopted a number of resolutions challenging the estimates of produce of land in the precanal and post-canal days and the Development Act. By the beginning of February the agriculturists of the canal areas were seriously affected, due to the enforcement of the Development Act. The government began harrassing poor cultivators for the realisation of the canal tax and made an effort to recover the arrears of taxes. On 10 February the cultivators' grievances were brought to the notice of the Burdwan Maharaja who suggested a public protest meeting be held to air the grievances of the ryots. On 14 February 1937, about 1,000 representatives of the cultivators of the Damodar Canal area attended a conference and resolved,

That in the opinion of the conference the principles underlying the Bengal Development Act and sections thereunder were arbitrarily opposed to the interests of the prajas and krishaks in general in the sense that they had been placed outside the jurisdiction of the civil court so that the appreciation of the Act might make the executive officers all powerful and give them absolute, arbitrary and unfettered authority which was sure to be used to oppress the ryots, that an estimate of the surplus produce of lands in the Damodar Canal area made by the officials of the irrigation Department was devoid of logic and was not based on facts, that the amount of paddy produced in the canal area did not admit of a taxable surplus after the deductions for payment of rent to the zamindar and expenditure on cultivations

The Canal tax agitation emphasised that since there had been no development and no increased amount of land situated in the Damodar Canal area, no improvement levy could be imposed under the Bengal Development Act 1935. The levy was linked to the government's extravagant capital expenditure, not to the paying capacity of the agriculturists or benefits, if any, derived by them. The improvement levy was thus totally illegal, unjust. unreasonable and contrary to the facts and opposed to justice. The peasants insisted that the cost of the works be recovered from the East Indian Railways, the Grand Trunk Road, Burdwan-Kawa Railway, Bengal-Nagpur Railway, the city of Calcutta and other vested interests who had really benefited from the canal. Further, they also clarified that since they derived benefits from the canal only once in seven or eight years when drought occurred, they should bear only that cost.

The Congress, which had also joined the agitation, set up its own inquiry committee. The report of the committee established that the Damodar Canal had not to any appreciable extent increased the productivity of the area served by it. The yield per acre prior to the construction of the canal was about 24 maunds and it remained at the same level after the canal had started functioning. The report also highlighted the fact that the 'Damodar Carnal project was never meant for irrigation purposes alone. It was intended "inter alia" that the canal should protect the railways, the G T Road, the Burdwan town, the port of Calcutta, etc. by modulating the strength of the Damodar flood. It recommended reduction of the tax on cultivators to reflect better the benefits against drought provided by the canal.

As a result, the government reduced the levy to Rs. 2-9-0) per acre. But even this amount was too high. In 1939, the government started attaching movable properties of the defaulters in the canal area. The local cultivators were determined to launch a satyagraha movement till the tax was further reduced to Rs. 1-80. per acre. The government on the other hand, sent a large contingent of Gorkha soldiers to bring the situation under control. Section 144 was promulgated, prohibiting public meetings. The 'policy of terrorisation' forced the peasantry to give up their agitation for a reduction in the canal tax. The people subsequently accepted the government rate of Rs. 2-9) per acre and paid the arrears.

LARGE DAMS

The capacity to divert rivers from their natural course increased dramatically in the post-colonial period with the transfer of technology of large dams from the US. The Bureau of Reclamation and the Army Corps of Engineers were in competition with each other and created a new culture of gigantism, financed by public money. In Reisner's words, 'what had begun as an emergency program to put the country back to work, to restore its sense of self-worth, to settle the refugees of the Dust Bowl, grew into a nature wrecking, money-eating monster that our leaders lacked the courage or ability to stop. Interest groups have mushroomed around the building of large dams, and their interests are in conflict with those of indigenous populations and ecologists. As Barnett has observed, Public water projects-dams, reservoirs, irrigation pipelines, schemes for rerouting streams and a host of other feats of engineering-are viewed by proponents-usually the U.S. Army Corps of Engineers who build them, politicians looking for federal money, and large agribusiness interests and power companies, which directly benefit from them-as modern wonders of the world. Opponents-environmentalists, Indian tribes whose land is taken or ruined, and cost-conscious politicians and bureaucrats denounce the implantation of huge concrete waterworks in the midst of America's wilderness and the accompanying hydraulic technologies as the worst examples of pork barrel politics and extravagant wasted

When the technological euphoria of dam building was transferred to India, the concomitants of ecological disruption and social conflicts were also transferred. These conflicts and destruction are more aggravated in India than the havoc caused in the American West because India is a riparian civilization which has evolved in a monsoon climate. Most of India's river valleys are highly populated and rivers have provided the primary life-support systems for our riparian settlements. Large dams, intensive irrigation and large diversions have, therefore, been associated with three types of conflicts. The first type is related to large-scale displacement and uprooting of people from their ancestral homelands leading to ecological refugees. This conflict, which originally expressed itself through human rights struggles based on the violation of rights of displaced people, has now taken an e cological turn, with human rights issues being perceived as intimately linked with ecological issues. The second type of conflict related to water projects arises from the ecological impact of impounding large quantities of water, transporting it across drainage boundaries and using it for intensive irrigation. Displaced people are, of course, in direct conflict with those who benefit from large dams and massive irrigation systems. However, when dams and canals cause waterlogging, even the 'beneficiaries' fight against state planned water projects.

Changes in water flows create changes upstream as well as downstream. Such changes generate conflicts not merely between the people and the state, but also between different communities and different states. The third type of conflict which is an outcome of large river diversions is regional conflict over water rights. Interests of people of different regions are articulated through regional governments, and regional conflicts take the form of inter-state conflicts over the sharing of river waters.

The Krishna river, one of the most important rivers of South India, was chosen for the ecological analysis of conflicts over river waters since it traverses through the most arid and drought prone regions in South India and there are intense and diverse demands for its water from different regions for diverse uses.

WATER CONFLICTS IN THE KRISHNA BASIN

The east flowing krishna river originates in the mahadev range of the western ghats, north of the hill station of Mahabaleshwar in Maharashtra, and after flowing over a distance of about 1,40(1 km it meets the Bay of Bengal, south of Vijayawada. In between its origin at 1,337 metres above the MSL and its delta, the river fows across the entire width of the Indian peninsula through the states of Maharashtra, Karnataka and Andhra Pradesh.

The Krishna river is joined in its course by a large number of tributaries, big and small, draining a total basin area of about 256,000 sq km of which the share of the three riparian states of Maharashtra, Karnataka and Andhra Pradesh is 26.8 per cent, 43.8 per cent and 29.4 per cent, respectively. The basin drains a length of approximately 700 km of the Western Ghats which is the predominant source of water of the river. As the river flows about 135 km from its origin near Mahabaleshwar Hills, it is joined by the Koyna river flowing from the western side of the same hill. Further along its course. it is joined by tributaries like Varna, Panchaganga and Dodhganga draining about 150 km of the Western Ghats. As the river emerges from the eastern slopes of the Western Ghats, it is joined by Ghataprabha and Malaprabha from the south at a distance of approximately 500 km from the origin.

After traversing the Deccan proper the east flowing Krishna then enters the alluvial lands and at a distance of about 800 km from the source, just before it enters Andhra Pradesh, a major tributary Bhima, draining the Western Ghats, north of Mahabaleshwar joins it from the north. Near Kurnool the river is joined by another major tributary, Tungabhadra from the south, draining a major section of the Western Ghats in Karnataka. Within a short distance from this confluence, the river enters the Naliamali Ranges characterised by peep gorges. At this place the Srisailam Dam and further downstream the Nagarjunasagar Dam have been constructed. At this point the major water sources of the Western Ghats have all been united.

Tributaries like Dindi, Musi, Palleru and Muneru draining the dry north-eastern parts of the basin join the river between Srisailam ad Vijayawada but do not add much water. Below Vijayawada, where the Krishna is blocked by the barrage constructed during the British period, the river spreads out into the delta and below the last major village Nagailanka it joins the Bay of Bengal in three branches, thus ending the long eastward journey of the waters of the Western Ghats.

Since water flow creates an interconnectedness within the basin, each intervention in land and water use, depending on its scale, can become the source of conflicts. The mining of iron ore at Kudremukh and Manganese ore in Sandur in the upper catchments of Tungabhadra has seriously affected the stability of the catchment and has led to severe soil erosion and silting of the Tungabhadra reservoir, thus conflicting with irrigation needs.

The Krishna river system has a large number of small, medium and major dams starting from Dhom which is located within 5 km of its origin. This storage and diversion of water from the original river course has destroyed the fishing economy which was dominant on both banks of the river as well as the indigenous irrigation system that existed throughout the course of the river. Further, large dams have also generated conflicts by creating waterlogging in the command areas. The hydroelectric power generation from the river water has come into conflict with irrigation needs both in terms of the spatial and temporal characteristics of water storage and distribution. The maximisation of power generation from Koyna demands that the water of the Krishna basin, draining into Bay of Bengal, be diverted to the Arabian Sea.

Industrial uses of the river system are a major source of conflict. For example, the pulp based industries on Tungabhadra have polluted the river and destroyed the fishing economy 20 km downstream. Moreover, the large-scale cultivation of pulpwood species like Eucalyptus in this part of the basin has impaired the groundwater recharge potential.

In the Krishna basin comprising mostly of arid and semi-arid regions water management had reached a high level of sophistication, both for surface as well as groundwater utilisation. An aerial view of the basin reveals a network of a large number of tanks, some pre-historic, others constructed by the local people or the rulers at different times in history. In general the technology used for all these tanks involved the construction of an earthen embankment at the exit of a natural water collection point that is a result of topography.

These tanks were used for surface irrigation of approximately 500 acres of land as well as for enhancing ground water recharge to support the wells. These tanks formed a network so that water did not drain out easily and was conserved at the site. To some extent indigenous water management techniques also included the diversion of streams to irrigate land by canals. The total number of tanks in the basin may be around 30,000. By arresting the scanty rainfall, these tanks actually provided a cushioning effect against variations in rainfall which is common in the basin. This decentralised water conservation system met both drinking water and agricultural needs. There was no major long distance transfer of water and the local cropping pattern evolved in accordance with the local water endowment.

The needs of the Vijayanagar Empire led to the first major intervention in the natural water flow. In the sixteenth century, specially during the reign of King Krishnadevaraya, there were many attempts to divert the water of Tungabhadra through seven canals in the Bellary district, these are now known as the Vijayanagar Canals. The canals provided water for irrigation as well as satisfied the needs of the large army stationed in the capital city of Hampi. The interests of the Vijayanagar rulers were not limited to canals. Understanding the crucial role of tanks in food production as well as in providing drinking water supply, the kingdom undertook a systematic programme of tank construction. The Daroji tank and the Vyasayaraya Samudram in Cuddapah district are the result of this programme.

The first large-scale intervention in the natural flow of water in the Krishna river basin was seen in the late nineteenth century. It was motivated both by the irrigation needs of export crops like cotton and groundnut, as well as for transporting these products easily to major ports like Madras.

The Krishna delta canal system based on the Vijayawada barrage was constructed in 1855 The Nira Canal in Maharashtra was constructed in 1835 to irrigate about 150,000 acres and the Kurnool Cuddapah Canal was constructed in 1886 to irrigate 100,000 acres. With the passage of time. an increasing number of government aided large and medium projects came up and today the Krishna river has numerous dams including the Dhom Dam which is at a distance of 5 km from its source. Midstream, we find the Alamatti and Narayanpur Dams of the Upper Krishna Project while further downstream Srisailam and Nagarjunsugar Dams generate electricity and divert water for irrigation.The tributaries have also been used extensively in this respect.

The Koyna Dam is situated 58 km below the origin of the river. The Tunga river is impounded at Gajanur and Bhadra at Lakavalli. The Tunga and Bhadra meet and the Tungabhadra Dam is located 265 km from the origin. In Ghataprabha the reservoir at Hidkal in Karnataka is the major irrigation project while Malaprabha is impounded at the peacock gorge near Manoli. The spread of water-intensive cultivation throughout the basin has dramatically altered the water balance, leading to major conflicts between water for cash crop cultivation and staple food production on the one hand, and between irrigation and drinking water needs on the other. The case of sugarcane cultivation in Maharashtra and grapes in Hyderabad are two instances of over-exploitation of water resources in the basin for cash crop production and a consequent destabilisation of the water cycle, leading to water scarcity in large parts of the basin.

Dams for irrigation and/or power are also a source of conflict between the traditional rights of people to land and water and the rights of the state to displace and uproot them for building river valley projects as in the case of Srisailam Dam. Large dams require massive submergence areas, and hence necessitate the displacement of large numbers of people. Big dams also allow large diversions of water. Major diversions from the river basin as in the case of the Telugu-Ganga Canal taking off from Srisailam Dam, affect the riparian rights of the states and have generated unresolvable inter-state conflicts.

DAMS AND DISPLACEMENT

The krishna, like other rivers of india has been reversed by the people srisailam is the most sacred pilgrim spot on the Krishna. It is named alter the Srisailam temple situated amidst rich forests on the banks of the river. The Krishna flows 3 km below the Srisailam temple which is dedicated to LOrd Shiva. In 1960. this ancient temple gave way to a temple of modern India Srisailam Dam.

The project 'the Srisailam project began in 1960, initially as a power project, across the Krishna, near Srisailam in Andhra Pradesh. After several delays, the main dam was finally completed twenty years later in 1981. In the meantime the project was converted into a multipurpose one with a generating capacity of 770 MWs by its second stage which was expected to be completed in 1987. The dam is to provide water for an estimated 4,95,000 acres with its catchment area of 79,553 sq miles and water spread of 238 sq miles. Under the right branch canal 1,95,000 acres in Kurnool and Cuddapah districts will have assured irrigation. From the initial modest estimate of Rs. 38.47 crores for a power project the total cost of the multipurpose project was estimated to cross Rs. 1,000 crores in its enlarged form. The 470 feet high and 1,680 feet wide dam has alone cost Rs. 404 crores together with the installation of four generating sets of 110 MWs each. The right branch canal is estimated to cost Rs. 449 crores and the initial investment of Rs. 140 crores has been provided by the World Bank. The projected cost-benefit ratio of the project has been worked out at 1:1.91 at 10 per cent interest on capital outlay.

The construction of the project has meant the submergence of 106,925 acres of land belonging to 117 villages (100 main and 17 hamlets). Of these villages, spread over six taluks of Kurnool and Mahaboobnagar districts. seventy-two were completely submerged and ten were partially submerged. A total of 27,871 families in these villages living in 21,080 dwellings had to be evacuated: resettlement had to be provided for nearly 158,00 people. In the summer of 1981, shocked by the brutal and inhuman manner in which people were thrown out of their homes by the government with the assistance of police, bulldozers and workers from the town, a Lokayan team in Andhra Pradesh carried out a survey of the problem of the evictees in July-August 1981. The survey aimed at:

1. Understanding the socio-economic background of the affected households.
2. Appraising government policies and programmes with regard to compensation and the rates paid, adequacy of the mode of payment and the reactions of the recipients.
3. Studying government actions in evacuation of the people facing displacement.
4. Examining the problems associated with the rehabilitation schemes.
5. The survey covered fifteen of the 100 villages which were affected, using a stratified random sample of 344 households.

A second survey was carried out in 1984-85 covering nine of the villages included in the earlier sample survey as part of the UN University project on Conflicts over Natural Resources. The questions that prompted the second survey were:

1. How are the people evicted from the submerged areas of the Srisailam Dam in May-June 1981 living now, i.e., what changes have taken place in the last three years?
2. How have they utilised the compensation paid to them by the government?
3. Who can do what to improve the conditions of this displaced people?
4. Given the experience of evictions under the Srisailam project, what measures should be undertaken for future projects to avoid such calamities?

A summary of the two reports highlighting the tragedy of the situation is presented here. It raises several important questions not only related to economics and development but to ethics as well.

SOCIO-ECONOMIC CONDITIONS OF THE PEOPLE BEFORE THE EVICTIONS

The soils of the river bank are very fertile and mostly black or red in colour. Farmers have been cultivating them for generations, if not centuries, growing a multiple variety of crops ranging from food crops like rice, jowar and other millets to cash crops like tobacco, chill), groundnut, vegetables, onions, mustard and wheat. The river bed was also cultivated in the dry season, especially by the weaker sections, harvesting a rich crop of water melons.

Overwhelmingly (81 per cent of the sample), the population in the region belongs to the weaker sections of society, i.e., the Scheduled Castes (14 per cent) and Backward Castes (67 per cent). As in other regions of the country, the Scheduled Castes are concentrated among the small and marginal farmers (66 per cent) and landless labourers (20 per cent). Backward Castes are predominant in all classes but are most numerous among those involved in non-agricultural occupations (84 per cent). Among the other castes

Reddys are predominant. People of other castes are more concentrated in the upper classes. They account for 42 per cent of the big farmers but only 5 per cent of the agricultural labourers. Of the total number of households, 17 per cent were those of big farmers, 16 per cent of middle farmers, 36 per cent of small and marginal farmers, and 17 per cent of agricultural labourers. We have classified those people as big farmers who owned more than 10 acres of wet lands' middle farmers as those owning between 5 to 9.99 acres, small as those owning between 2.5 to 4.99 acres, marginal as those owning between 0.1 to 2.49 acres. Two acres of dry land has been assumed to be equivalent to 1 acre of wet land based on the income generated.

In addition to agriculture, small and marginal farmers are also involved in several subsidiary occupations such as sheep and goat rearing, toddy tapping, weaving, fishing and plying dinghies across the river. Those belonging to service caste groups like barbers and washermen tried to supplement their income from their caste occupation by working on land either as small or tenant farmers or agricultural labourers. The proportion of agricultural labourers is comparatively low in these villages, because the poorer sections of the population are engaged in the cultivation of poromboke and manyam (waste) lands.

The average size of the displaced family was found to be around seven in both sample surveys (census figure: 5.33). One interesting feature was that big farmer households had an average size of ten members per family, perhaps there were more joint families among them.

Draught power provided by bullocks appeared to be adequate in these villages. In addition to cows and buffaloes, a majority of households also reared sheep, goats and fowls. In view of the prosperous agriculture in these areas, employment prospects were particularly good, in the sense that people could secure employ ment for approximately 250 days in a year. In addition, there was immigrant labour from neighbouring areas during the peak seasons. The average annual income per household was approximately Rs. 8,000 with a minimum of Rs. 2,000 and a maximum of Rs. 150,000.

Since stone is available in plenty and owing to the relative prosperity of the region, most of the houses, including those of the poor and landless (81 per cent in the second sample), were made of stone and were quite spacious, although they were old.

From the data it is evident that though the region is relatively prosperous, it is not very different from other parts of rural India in terms of complacency regarding caste and class. Economic power is largely concentrated in the hands of the 'other castes' (Reddys in this region). The leaders in these villages are drawn from the 'other castes' (Reddys and Velamas), and they exercise tremendous influence on the people in the village. They also maintain close links with government officials in towns and with other important individuals. Often villages are divided into various factional groups following One leader or the other. All these factors had their respective impact on the entire process of displacement and rehabilitation.

One important factor that needs to be noted here is that for one generation, i.e., twenty years (1960 to 1980, i.e., till their eviction) no significant developmental activities were undertaken in the submergible areas since logically the whole area would be under water 'very soon'. The people of this region therefore had to do without electricity, proper roads, school buildings and other government asset building activities.

COMPENSATION

Keeping in view some of the problems encountered in trying to rehabilitate people displaced by developmental projects, the g,overnment of Andhra. Pradesh decided to pay compensation in cash, a policy initiated with the Pochampad project in the late sixties. Compensation was assessed for lands, wells and houses. Acquisition began in July 1969 and 1,829 acres of land was acquired. But this process was stopped almost immediately due to non-availability of funds and the areas notified for acquisition were once again denotified. Acquisition was resumed in 1974 and completed by 1980, in accordance with the Land Acquisition Act, 1894 (and several resolutions adopted and recommendations made at meetings of the Srisailam Control Board and the government). Compensation was paid for 84,772.55 acres out of a total of 107,348 acres submerged. The rest of the land was either government wasteland, or government land assigned to the poor or forest land constituting 20 per cent of the total submerged area which was not compensated for.

Drawing upon the registered value of sale/purchase of land which is normally much lower than the actual price (to avoid higher stamp duty) and the fact that the acquisition and payment of money were long drawn out affairs, they had the net effect of offering very low prices for the properties acquired. The average price paid for dry lands was Rs. 1,820 per acre, whereas the prevailing market rate for reasonably good dry land was Rs. 10,000 per acre in 1981. Similarly, compensation paid for wet land was on an average Rs. 3,547.05 per acre while the market value was around Rs. 20,000 per acre, i.e., the compensation paid between 1974 and 1980 was only one-fifth of the market value of land at the time of eviction.

BIBLIOGRAPHY

Barro, Robert J., : *Economic Growth.* McGraw-Hill, 1995.

Bhagwati, Jagdish: *India in Transition: Freeing the Economy,* Oxford, Clarendon Press, 1993.

Chandhok, H.L.: *The Economy: Annual Time Series Data,* New Delhi, Living Media India, 1990.

Dahiya, L.N.: *Dynamics of Economic Life in Rural India,* Delhi, Gian, 1991.

Ehrenberg, Ronald: *Modern Labor Economics,* Harper Collins, 1994.

Greene, William H.: *Econometric Analysis,* MacMillan Publishing Company, 1993.

J.H. Dully, and R. Jolly: *Redistribution with Growth,* London, Oxford University Press, 2001.

Kelly, F. P.: *Charging and Accounting for Bursty Connections,* Massachusetts, MIT Press, 1997.

Little, I.M.D.: *India: Macroeconomics and Political Economy, 1964-1991,* Washington, World Bank, 1994.

Mongia, J.N.: *India's Economic Development Strategies, 1951-2000 A.D.*. New Delhi, Allied, 1986.

Petre, D., & Harrington, D.: *The Clever Country? Australia's Digital Future,* Sydney, Lansdowne Publishing Ltd, 1996.

INDEX

□□□